Discover Probability

How to Use It, How to Avoid Misusing It,
and How It Affects Every Aspect of Your Life

Discover Probability

Probability

**How to Use It, How to Avoid Misusing It,
and How It Affects Every Aspect of Your Life**

Arieh Ben-Naim

The Hebrew University of Jerusalem, Israel

World Scientific

NEW JERSEY · LONDON · SINGAPORE · BEIJING · SHANGHAI · HONG KONG · TAIPEI · CHENNAI

Published by

World Scientific Publishing Co. Pte. Ltd.

5 Toh Tuck Link, Singapore 596224

USA office: 27 Warren Street, Suite 401-402, Hackensack, NJ 07601

UK office: 57 Shelton Street, Covent Garden, London WC2H 9HE

Library of Congress Cataloging-in-Publication Data
Ben-Naim, Arieh, 1934–   author.
   Discover probability : how to use it, how to avoid misusing it, and how it affects every aspect of
your life / Arieh Ben-Naim, The Hebrew University of Jerusalem, Israel.
      pages cm
   Includes bibliographical references and index.
   ISBN 978-9814616317 (hardcover : alk. paper) -- ISBN 9814616311 (hardcover : alk. paper) --
ISBN 978-9814616324 (pbk. : alk. paper) -- ISBN 981461632X (pbk. : alk. paper)
   1. Probabilities. I. Title.
   QA273.B445 2014
   519.2--dc23
                                                        2014032704

British Library Cataloguing-in-Publication Data
A catalogue record for this book is available from the British Library.

Printed in Singapore by Mainland Press Pte Ltd.

This book is dedicated
to you,
the reader who is holding this book
at this moment and pondering:
To read or not to read?

Haman the villain, casting a lot

3:6 And he thought scorn to lay hands on Mordecai alone; for they had showed him the people of Mordecai: wherefore Haman sought to destroy all the Jews that were throughout the whole kingdom of Ahasuerus, even the people of Mordecai.

3:7 In the first month, that is, the month Nisan, in the twelfth year of king Ahasuerus, they cast Pur, that is, the lot, before Haman from day to day, and from month to month, to the twelfth month, that is, the month Adar.

Megillat Esther (Book of Esther, 3:6,3:7))

(ו) וַיִּבֶז בְּעֵינָיו לִשְׁלֹחַ יָד בְּמָרְדֳּכַי לְבַדּוֹ כִּי הִגִּידוּ לוֹ אֶת עַם מָרְדֳּכָי וַיְבַקֵּשׁ הָמָן לְהַשְׁמִיד אֶת כָּל הַיְּהוּדִים אֲשֶׁר בְּכָל מַלְכוּת אֲחַשְׁוֵרוֹשׁ עַם מָרְדֳּכָי: (ז) בַּחֹדֶשׁ הָרִאשׁוֹן הוּא חֹדֶשׁ נִיסָן בִּשְׁנַת שְׁתֵּים עֶשְׂרֵה לַמֶּלֶךְ אֲחַשְׁוֵרוֹשׁ הִפִּיל פּוּר הוּא הַגּוֹרָל לִפְנֵי הָמָן מִיּוֹם לְיוֹם וּמֵחֹדֶשׁ לְחֹדֶשׁ שְׁנֵים עָשָׂר הוּא חֹדֶשׁ אֲדָר: **מגילת אסתר (ג, ו-ז)**

עַל כֵּן קָרְאוּ לַיָּמִים הָאֵלֶּה פוּרִים עַל שֵׁם הַפּוּר
מגילת אסתר [ג, כו]:

Preface

If you are someone who is curious about the "laws" that govern all of the events that we observe in our lives every day, this book is for you. I have thoughtfully conceptualized, designed and written this book with the hope of satisfying your curiosity, as well as nourishing your desire and quest for knowledge. You do not need to know anything about mathematics, nor are there any other prerequisites to reading and understanding this book. All you need is simple common sense, and a strong determination to use it.

I have invested a great deal of effort in making this book readable and enjoyable. As you leaf through the pages of the book, you will be led and guided towards understanding the "laws" of probability. As you discover and immerse yourself in these "laws," you will learn how to make probabilistic judgments. You will also learn how to avoid misusing probabilistic arguments, how to be critical and how not to fall into the pitfalls that are created by some authors who abuse probability in trying to convince you of false propositions.

You will notice that I have enclosed the word "laws" in inverted commas in order to emphasize that I am not going to discuss any *law* of physics, chemistry or biology. Yet these "laws" of probability are involved in any branch of science, as well as in any aspect of our lives. These laws are sometimes referred to as the "laws of probability," the "laws of randomness" or the "laws of regularities and predictability" in the outcomes of experiments that are inherently unpredictable and irregular. In other words, the theory of probability is about "order in the disordered world." As in my previous book, I have designed a series of experiments that you can carry out with either coins, dice or marbles or simulated experiments on a computer, or you can simply imagine that you are doing the experiment.

To acquire familiarity with the fundamentals of probability theory, you do not need to know any mathematics. Of course, the theory of probability as a branch of mathematics involves highly sophisticated mathematics. However, one does not need any mathematical knowledge in order to appreciate the meaning of probability, the basic laws of probability and some simple usages of probability. It will help if you can do some of the simulations on a computer, but if you cannot, you can imagine doing the simulation, and accept the results that I present. To help you "digest" these results, I will also provide you with some plausible arguments.

The term "probability" is used in two broad senses. The first use is when it is applied to propositions, statements, assertions, conjectures, theorems, etc. For instance:

(a) This book was written by Shakespeare.
(b) The number of stars is larger than 10^{10}.

One can ask: "What is the probability that statement (a) is true? What is the probability that statement (b) is true?"

The second use of probabilities is in estimating the likelihood of the occurrence of an *event*, or an outcome of an experiment.

For instance:

(a) Tomorrow it will rain.
(b) You will win the grand prize of the lottery during this year.

Here, we are asking about the likelihood that it will rain tomorrow, or that you will win a prize.[1]

Thus, in the first case, one asks about the likelihood that a statement or a proposition is true or false. In the second case, one asks about the likelihood that some event will or will not occur.

The line between propositions and events is not always clear-cut. Sometimes, one can translate an *event* into the language of *proposition*, and vice versa. In most of the sections of this book, we shall use the term "probability" in this second sense.

Consider this: In the Yahoo weather forecast for tomorrow, you read:

Precipitation: 40%
Relative humidity: 50%

Do you know what these numbers mean? Write down what you think they mean before looking at Note 2.

As I mentioned above, this book does not rely on any knowledge of mathematics. By saying that, I mean that I do not use any sophisticated mathematical theorems. This does not mean, however, that I shall not use some mathematical notations and simple mathematics. The complete list of the mathematical notations used in this book is collected in Note 3.

The book is organized into eight sessions. In Session 1, we start with the meaning of probability. I shall first try to convince you that you already have a rudimentary sense of probability. For very simple cases, this "sense of probability" is sufficient to make a prudent decision. However, in more complicated cases, intuition often fails us. We shall learn how to sharpen our intuition and hone our skills in solving simple probabilistic problems. Some of the problems are purely recreational, but some can be matters of life and death.

In Session 2, we shall discuss two "definitions" of probability. These are "methods of calculating probabilities" rather than definition. Session 3 is devoted to the so-called axiomatic approach to probability. This is the *foundation* on which the whole theory of probability is erected. In Sessions 4 and 5, we discuss *dependence* and *independence* between events, and the concept of *conditional* probability. The latter concept is central to probability theory and is responsible for turning probability theory into a unique and fascinating topic. In Session 6, we briefly discuss the concept of averages and standard deviations. Session 7 presents some of the most important probability *distributions*.

In Session 8, we arrive at the culmination of the entire book, where we discuss the elements of information theory. Although this topic is not usually discussed in textbooks of probability, it is a *purely probabilistic concept*, full of meaning and plentiful applications, and what is more surprising is that this theory resolves one of the most mysterious and difficult to understand quantities in science — entropy.

Every day of our lives, we make decisions between two or more options, the results of which are not certain. Which method of transportation will be faster, which will be safer, which will be more convenient? For all of these questions, we have to make a probabilistic judgment. Sometimes our intuition helps us, but sometimes it fails us.

The artistic renditions throughout the book were created by my friend, Alex Vaisman. They are part of the text and they all convey some probabilistic message. The reader is invited to "invent" probabilistic questions for the scenes shown in these drawings.

By reading this book, you should learn to be able to make probabilistic assessments prudently, and also avoid some common misuses of probability. However, I cannot promise you foolproof protection from falling into one of the probabilistic traps or pitfalls that are used — either consciously or unconsciously — by some authors.

I hope that the use of mathematical notations will not dissuade you from reading this book. On the contrary, I hope that the reader who is unfamiliar with some of these notations will find them useful and a rather efficient way of writing, instead of using and repeating long sentences describing the operation in which the notation is used.

Finally, I urge you to read this book *actively* and *critically*. You should do as many of the exercises as possible. Try to solve the exercises, and if you cannot, look at the notes for hints and solutions. Always be alert and critical, and do not take anything that I write for granted. If you have any questions or comments, or even criticisms, please feel free to write to me (ariehbennaim@gmail.com), and I promise to do my best to help.

Acknowledgements

I am grateful to Ruma Falk for sending me some of her articles on children's perceptions of probability. I also enjoyed reading her books on probability, from which I took a few examples. I thank Robert Aumann for providing the full story behind the probabilistic choices pilots were confronted with during WWII.

I am very grateful to Alexander Vaisman for providing the beautiful illustrations throughout this book.

I am also indebted to many of my friends and colleagues — Max Berkowitz, Frank Bierbrauer, Claude Dufour, Steven van Enk, Zvi Kirson, Azriel Levy, Xiao Liu, Thierry Lorho, Robert Mazo and Erik Szabo — for reading parts or the entire manuscript and offering helpful comments. I am also indebted to the artist Chuan Ming Loo for the beautiful cover design of the book. It is also a pleasure to acknowledge the help and friendly correspondence with the editor Sook Cheng Lim.

As always, I am most grateful to my wife, Roberta, who helped me in each and every stage of writing this book, from typing to editing to re-editing, and her suggestions to improve the clarity of the book.

Contents

Session 1

What is Probability?

Before we start the first session of this book, I suggest that you go to a library and look at some books on probability. Open a few books and look in the index for "probability, definition." After browsing through the first few pages of these books, you will find that some books do not *define* the term probability. Others provide some kind of a "definition," but if you read carefully, you will realize that the definition of the term "probability" uses the concept of probability. In other words, these definitions are circular, i.e. the definition of probability is based on the concept of probability or another equivalent concept, such as randomness, likelihood, etc. Some books introduce the concept of probability *axiomatically*, which means that a number is attached to every *event*, and this number fulfills some properties. We shall discuss these properties in Session 3. This is, of course, not a definition of the term "probability." Other books might suggest a "definition" of probability by providing a few examples of how to *calculate* the probabilities of certain events (we shall do this in Session 2).

In addition, I suggest that you read the first paragraph in Wikipedia on "probability." You will find the following (as of

1

January 2013 — this might change with time, as Wikipedia's contents are constantly evolving).

Probability (or ***likelihood***) *is a measure or estimation of how likely it is that something will happen or that a statement is true. Probabilities are given a value between 0 (0% chance, or will certainly not happen) and 1 (100% chance, or will certainly happen). The higher the probability, the more likely the event is to happen, or, in a longer series of samples, the greater the number of times such event is expected to happen.*

Here are some more "definitions" taken from an encyclopedia: *Probability Definition*

(1) *The quality or state of being probable; appearance of reality or truth; reasonable ground of presumption; likelihood.*
(2) *That which is or appears probable; anything that has the appearance of reality or truth.*
(3) *The chance that a given event will occur.*
(4) *Likelihood of the occurrence of any event in the doctrine of chances, or the ratio of the number of favorable chances to the whole number of chances, favorable and unfavorable.*

Before you continue, please read again the "definitions" above and pause to think and write a short statement describing what you have learned from this paragraph. Do you find any of these "definitions" satisfactory?[1]

In this session, I want to convince you that, in spite of the fact that no definition of probability exists, you already know what probability is. You have a sense of what it means when one says that the probability of an event A is larger or smaller than the probability of an event B.[2]

This sense of probability is much the same as the sense of beauty you have without having a definition of "beauty." You can confidently say that one landscape is more beautiful than another. This example should not mislead you into concluding that the sense of probability is always a subjective sense. In some cases it is subjective, but in most cases when it is used in the sciences, it is an objective measure — objective in the sense that everyone who is using these measures will agree with them.

Read the following two statements and assess their objectivity or subjectivity.

(i) If you marry this lady, it is very probable that you will be happy all your life.

(ii) If you bet on the outcome "4" in the throwing of a fair dice, it is probable that you will win on average in one out of six throws of the dice.[3]

It is not only you, the reader of this book, who has a sense of probability, or a sense of likelihood, even though you lack a definition for probability. Young children also have a sense of probability, even though they have never heard of the word "probability." The following section should convince you of the validity of the above statement.

1.1 Children's Perception of Probability

During the early 1970s, I heard a lecture given by Ruma Falk on some research she had carried out on "how children perceive the concept of probability." The reported results were fascinating and full of surprises.[4]

Since then, many researchers have examined how young children perceive the concept of probability. I will show you only a few examples of experiments conducted on children aged 4 to 12. The experiment described here is a variation of a similar experiment carried out by Falk *et al.* (1980 and 2012).

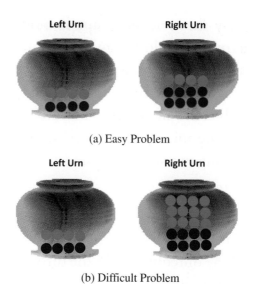

Left Urn   Right Urn

(a) Easy Problem

Left Urn   Right Urn

(b) Difficult Problem

Fig. 1.1 Two experiments with marbles in urns.

In the original publication, you will find many variations of these experiments and many thoughtful and insightful conclusions. I urge you to take a look at some of the articles listed in the bibliography.

Figure 1.1a shows two games that are denoted as "easy" and "difficult." The "easy" game consists of two urns. The one on the left contains four red and four blue marbles. The second urn on the right contains four red and eight blue marbles.

Children aged 4–11 were shown the two urns and their contents. They knew the contents of the left and the right urns. After examining the contents of the two urns, they were told to close their eyes, choose an urn and draw a marble from the urn they had chosen. They were also told that if the marble they drew was blue, they would be rewarded with a prize (win), while if they drew a red marble, they would get nothing (lose).

Which urn do you think the children chose? Which urn would you choose if you were the child asked to draw a marble

from the urns? Why did you choose that particular urn? Suppose
you chose the urn on the left, and did not win — what urn will
you choose the next time you play?

What is the probability of winning in this particular game
if you choose the urn on the right?

What is the probability of not winning if you choose the
urn on the left?

Write down your answers before you consult Note 5.

Most young children chose the urn on the right, which is the
correct choice, but for the *wrong* reasons. Furthermore, young
children who chose the urn on the right and did not win switched
to the urn on the left on the next game. When asked why they
changed their choice of urn, they simply said: "The first choice
was not good, it did not deliver the expected prize."

Figure 1.2 shows some of the results from Falk *et al.* (1980).

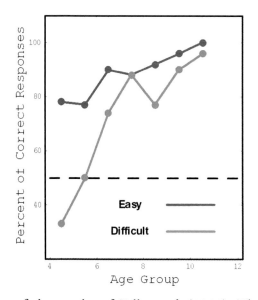

Fig. 1.2 Some of the results of Falk *et al.* (1980). The percentage of
children giving the correct answer, at different ages.

Older children, say 9 years old and above, chose the urn on the right on the first game. They continued to choose the same urn on the second and third games, even if they did not win on some of these games. Somehow, they sensed that even if the urn on the right did not deliver the prize, this urn was still the better choice. *Better* does not guarantee that one wins every game. It means that, on average, you will win with higher probability. Therefore, if you choose the urn on the right hand side and do not win, you should not be discouraged. Be patient, continue to play and stick to the same urn. In the long run, you will be winning.

Now for the likelihood of winning or losing:

For the urn on the left:

$$Pr(win) = \frac{1}{2}, \quad Pr(lose) = \frac{1}{2}.$$

For the urn on the right:

$$Pr(win) = \frac{8}{12} = \frac{2}{3}, \quad Pr(lose) = \frac{4}{12} = \frac{1}{3}.$$

Here, *Pr* is a shorthand notation for "probability of." Thus, *Pr(win)* means the probability of winning. The actual values given above should be understood intuitively. We shall discuss how to calculate probabilities in the next session.

You can see that the urn on the right has a higher probability or likelihood of winning, and therefore it is advantageous to choose this urn in this "easy" game. It is advantageous not because this urn has *more* blue marbles, but because the *ratio* of the number of blue marbles to red marbles is larger in the urn on the right than in the urn on the left.[6] From now on, we shall use the notation *Pr* as a shorthand notation for "probability," but you can always interpret probability as a measure of likelihood.

When considering the urn on the right, children do not *calculate* the probabilities of winning or not winning. Instead, they sense the ratio of the probabilities, or the ratio of the "odds" of winning and losing. Clearly, for the urn on the right, the ratio is:

$$\frac{Pr(win)}{Pr(lose)} = \frac{8}{4} = 2.$$

Whereas for the urn on the left, the ratio is:

$$\frac{Pr(win)}{Pr(lose)} = \frac{4}{4} = 1.$$

Thus, it is clear that the urn on the right is a better choice. The ratio of the probabilities is twice as large as that of the left urn. If we want to assign probabilities to "win" and to "lose," we must choose a *scale*. This simply means that we have to decide on the *range* of values that probabilities may have. We normally choose a scale between zero and one to express probabilities (more on this in Sessions 2 and 3). Here, we choose the probability of the event "either win or lose" to be 1, i.e. we choose our *scale* of probabilities such that the *certain* event "either win or lose" is 1. Hence, we write:

$$Pr(win) + Pr(lose) = 1.$$

Using the above ratio for the urn on the right, we get $Pr(lose) = \frac{4}{8}Pr(win)$. Hence, substituting $Pr(lose)$ into the last equation, we get:

$$Pr(win) + Pr(win) \times \frac{4}{8} = 1.$$

Hence, we can solve for the two probabilities for the urn on the right:

$$Pr(win) = \frac{2}{3}, \quad Pr(lose) = \frac{1}{3}$$

which we already had guessed intuitively.

Sometimes, we use percentages (%) to express probabilities. In this case, the sum of the two probabilities will be:

$$Pr(win) + Pr(lose) = 100\%.$$

Hence we get:

$$Pr(win) = \frac{2}{3} \times 100 \approx 66.67\%$$

$$Pr(lose) = \frac{1}{3} \times 100 \approx 33.33\%.$$

Of course, in daily life, we do not have such a neat way of calculating probabilities.

If you are going to travel tomorrow, and you have to decide whether or not to take an umbrella, you might estimate the likelihood of it raining (by whatever means you have), or you might have heard or watched the forecast that the probability of it raining tomorrow is 80%, or that the ratio of the probabilities is $\frac{Pr(rain\ tomorrow)}{Pr(no\ rain\ tomorrow)} = \frac{80\%}{20\%} = 4$.

If this is the case, you had better take the umbrella. This is not a precise way of calculating probabilities, but it is sufficient information to help us make a decision.

After contemplating the odds in the easy game shown in Fig. 1.1a, let us go to the more "difficult" game shown in Fig. 1.1b. This game is not really difficult, but within the research on children's perceptions of probabilities, it was considered to be the more difficult one.

In this game, the urn on the left has the same number of blue and red marbles (4 and 4) as in the easy game in Fig. 1.1a. The urn on the right contains 8 blue marbles and 12 red marbles (Fig. 1.1b).

Children who were exposed to this game responded differently according to their ages (see Fig. 1.2). The younger children made wrong choices most of the time. They chose the urn on the right. When asked why they had made these choices, they answered, "Because there are *more* blue marbles in the urn on the right than in the urn on the left." Clearly, these children had no sense of probability. As mentioned in connection with the easy game in Fig. 1.1a, young children chose the urn according to the *absolute* number of blue marbles; the more blues, the more attractive the urn. In the easy game, more blues coincided with a larger probability of winning. Therefore, the children chose the urn on the right for the wrong reason. In the more difficult game, the young children again chose the urn on the right. Here, they chose the wrong urn for the wrong reason — the absolute number of blue marbles.

Children aged 10 and above made better choices. Their sense of probability told them that the important quantity is not the absolute number of blue marbles (8 on right versus 4 on the left), but rather the *ratio* of the numbers of blue and red marbles.

For the game in Fig. 1.1b, the ratios are:

For the urn on the left:

$$\frac{Pr(win)}{Pr(lose)} = \frac{4}{4} = 1.$$

For the urn on the right:

$$\frac{Pr(win)}{Pr(lose)} = \frac{8}{12} = \frac{2}{3}.$$

Clearly, the better choice in this case is the urn on the left.
Now, pause and think:

Calculate the probability of winning if you choose the urn on the right and if you choose the urn on the left.

Suppose you chose the urn on the left, and you drew a red marble — would you change your choice the next time you play the game? Note that by playing the game a "second time," we mean *exactly* the same game. This means that whichever marble you draw in the first trial, it is returned to the urn before you play the next game, and so on. In Session 4, we shall discuss a slightly different sequence of games where the marble that is drawn is *not* returned to the urn. This variation of the game involves *conditional probabilities*, which we shall learn about in Session 4.

Now that you are convinced that not-so-young children have a sense of probability, let us test your sense of probability.

Consider the following simple games:

I throw a fair dice. By a "fair dice" I mean that it is a perfect cube, and you do not have any reason to suspect that one side is heavier than another. The faces of the cube are marked with different numbers of dots from one to six.

You see that the dice in the air whirls before it lands on the floor (see Fig. 1.3).

Fig. 1.3 A dice whirls in the air.

Game A

Choose a number between one and six, say "4." If the dice falls with the upper face showing the number 4 (we shall say for convenience that the outcome is "4"), then you get a dollar. If the outcome is different from "4," you get nothing.

Which number will you choose in the game?

Why did you choose this particular number?

How much do you expect to gain if you play this game 1000 times?

Game B

This is the same as in Game A, i.e. if the outcome is the same as the number you chose (say "4"), then you get a dollar. But if the outcome is different, you have to *pay* 21 cents.

Presuming you want to maximize your earnings, which number would you choose in this game?

How much do you expect to win (or lose) if you play this game 1000 times?

Answer these questions before you continue — they are easy questions. After answering, compare your results with Note 7.

Now consider a slightly more complicated game:

You are shown an urn (see Fig. 1.4) containing 4 blue marbles, 6 red marbles and 10 green marbles.

Fig. 1.4   An urn containing four blue, six red and ten green marbles.

You know the contents of the urn. You have to choose a *color* (not an urn as in the previous games) — either blue, red or green. Then you close your eyes and draw one marble. If you choose and draw a blue one, you get $2, but you do not get anything if it is not a blue marble.

If you choose a red marble and draw a red one, you get $3, but you do not get anything if it is not a red marble.

If you choose a green marble and draw a green one, you get $1, but you do not get anything if it is not a green marble.

Assuming you want to maximize your earnings, which color will you choose? Explain why you chose that particular color.

In order to answer the question above, you have to first calculate the following probabilities:

What is the probability of drawing a blue marble?

What is the probability of drawing a red marble?

What is the probability of drawing a green marble?

After calculating these probabilities, will you change the "color" of your choice?

The probability of drawing a blue marble is: $\frac{4}{20} = \frac{1}{5}$.

The probability of drawing a red marble is: $\frac{6}{20} = \frac{3}{10}$.

The probability of drawing a green marble is: $\frac{10}{20} = \frac{1}{2}$.

Clearly, the probability of drawing a green marble is the largest, simply because there are more green marbles. The probability of drawing a blue marble is the lowest, simply because there are fewer blue marbles.

When calculating your expected earnings, suppose you play the game 1000 times. Each time you play, you choose the same color, and after you draw a marble and get whatever you earned, you return the marble to the urn, and draw a marble again from the same urn under the same *initial* conditions. (In Session 4, we shall learn how to calculate *conditional* probabilities, i.e. situations in which you draw a marble but do not return it to the urn. In this modified game, the initial conditions are changed every time you play the game, see below.)

If you chose "blue," your average expected earnings are $2 \times \frac{1}{5} \times 1000 = \$400$.

If you chose a "red," your average expected earnings are $3 \times \frac{3}{10} \times 1000 = \$900$.

If you chose a "green," your average expected earnings are $1 \times \frac{1}{2} \times 1000 = \$500$.

Now you see that, in this game, the better choice is red. Although it has a smaller probability of being drawn compared with green, the average earnings are highest when the red marble is chosen.

If you chose red, then you should stick with this color. However, if you chose green because there is a higher probability of picking a green marble, then you should switch to red, which has higher expected earnings.

Now consider a more challenging problem. Suppose you chose red in the previous game, and suppose you drew a red marble and earned $3. Well done!

Now you repeat the game but with one difference: You do not return the red marble to the urn. Which color will you choose next? Note that now the initial conditions have changed (see Fig. 1.5). There are now 4 blue, 5 red and 10 green marbles. Do your own calculations before you look at Note 8.

Let us try the calculation for the next game. Suppose that in the second game you drew a red marble, and removed it from the urn. Which color would you choose now? Note that now the initial conditions have changed again (see Fig. 1.6);

Fig. 1.5 An urn containing four blue, five red and ten green marbles.

Fig. 1.6 An urn containing four blue, four red and ten green marbles.

there are now 4 blue, 4 red and 10 green marbles. Do your own calculations before consulting Note 9.

Again, you would be best off choosing red. However, be careful in the next step. If you drew red in the previous game, and did not return it, the urn now contains a total of 17 marbles, and the expected earnings are:

For "blue": $Pr = \frac{4}{17}$, with expected earnings of $\frac{4}{17} \times 2 = \frac{8}{17}$.

For "red": $Pr = \frac{3}{17}$, with expected earnings of $\frac{3}{17} \times 3 = \frac{9}{17}$.

For "green": $Pr = \frac{10}{17}$, with expected earnings of $\frac{10}{17} \times 1 = \frac{10}{17}$.

At this stage, you can see that it is better for you to switch to green. Although the prize for red is bigger, the probability of green is now much higher, such that the expected earnings are bigger for the choice of green.

I hope you did all the proposed calculations and compared them with the notes to this session. You will realize that although we did not *define* what probability is, you were able to calculate some simple probabilities by relying on your sense of probability. In fact, you have also calculated some *average* quantities, even though we did not define the term "average." We shall return to this term in Session 6.

Before we get into more details on the theory of probability, you should also be aware of the fact that the concept of "probability" is used in daily life as a measure of the extent of our belief in the occurrence of an *event*, and sometimes in the extent of truthfulness of certain *statements*, or propositions (such as the probability that this book was written by Shakespeare being 10%, or the probability that God exists being 80%).

In this book, we shall deal with the term "probability" as it is used in the sciences. Either we have a "recipe" on how to calculate probabilities (see Session 2) or we have collected statistical data from which we extract probabilities of various events.

It is not uncommon to find people misusing or even abusing the concept of probabilities. Before ending this session, I will write down eight statements. You should think about the meaningfulness of these statements:

(a) The probability of a book is 0.3.
(b) The probability of God is 0.9.
(c) The probability that Kukuriku exists is 0.6.
(d) The probability that the book is interesting is 0.8.
(e) The probability of it raining tomorrow is 0.8.
(f) The probability of you living until the age of 120 is 0.01.

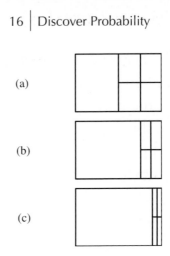

(a)

(b)

(c)

Fig. 1.7 Three boars each divided into five regions.

(g) The probability of obtaining an *even* result in throwing a fair dice is 1/2.

(h) The probability of obtaining the outcome "4" in throwing a fair dice is 1/6.

What is the meaning of all these probabilities?[10]

Exercise (a): You are shown three boards as shown in Fig. 1.7. Each board is divided into five regions. I throw a dart randomly at each of the boards, and you have to guess in which region the dart will hit. If you guess correctly, you win a prize — otherwise you do not get anything.

(1) Which *region* will you guess for each of the boards (i.e. which region on each board has the highest probability of being hit by the dart)?

(2) Which *board* will you choose if you have to play the same game but only once with the board of your choice?

Exercise (b): The same game as in Exercise (a), but you have to determine in which region the dart landed by asking binary questions, i.e. you ask questions that can be answered with either

YES or NO. You have to pay $1 for each question you ask. If you find out where the dart landed, you get $10.

(1) What strategy will you choose in asking your questions, assuming you want to maximize your earnings for each board?
(2) Which board will you choose if you have to play this game many times, but with only one board, such that you maximize your earnings?
(3) Once you choose a board and play the same game 1000 times, can you estimate how much you will earn on average?[11]

1.2 Casting Lots in the Present and in the Past

Nowadays, the practice of casting lots is used whenever we are forced to make a decision between two nearly equivalent choices. We believe that this method of making decisions is the "fairest" one. For instance, a man with two sons wrote in his will that each of them would get a painting each, A or B. He did not specify, however, which painting would go to his eldest son, and which one will go to the younger son. Upon his death, both sons happened to want painting A. Inasmuch as the will did not specify which painting was to go the each of the sons, both of them had equal rights to obtain said painting A. How can we decide who gets this painting? In such cases, it seems that the best or the fairest way of deciding who gets what is to toss a coin, and agree that if the outcome is a head, the eldest son gets painting A, whereas if the outcome is a tail, the younger son gets painting A.

In ancient times, people believed that the occurrence of events was ultimately controlled by deities, not by chance as we perceive things today. Thus, by casting lots, ancient peoples sought divine intervention. This practice of soliciting divine intervention is called *divination*. The Latin word for guessing is *adivinare*, and in Spanish it is *adivinar*, which has the same root-word of "divine," reminiscent of the belief in the intervention of a divine power in casting lots.

The bible contains several stories about people casting lots in order to make decisions. One of the most famous is that of Haman, the villain who cast the "Pur" (the lot or the "fate") in order to determine the date of annihilating the Jewish people during the time of the ancient Persian Empire. Clearly, Haman sought to consult a divine power in order to find the best day to destroy the Jews.

The Jewish holiday of "Purim" (the plural of "Pur," or lot, in Hebrew) commemorates the deliverance of the Jews from

עַל כֵּן קָרְאוּ לַיָּמִים הָאֵלֶּה פוּרִים עַל שֵׁם הַפּוּר. מְגִילַת אֶסְתֵּר [ג, כו]

Fig. 1.8 Haman casting the "Pur."

Haman's evil plot to destroy them. Jews celebrate the festive holiday of "Purim" on the fourteenth day of the Hebrew month of "Adar."[12]

1.3 Personal Interpretations of Probability

In our daily lives, we make many decisions based on probabilities. In most cases, the decisions involve many personal, subjective and biased arguments in addition to the "hard" facts. When a doctor tells you that surgery has a better chance of success than drug treatment, your choice depends on many other factors aside from the "dry" facts of statistics. You might choose the drug treatment because you are terrified of undergoing surgery, or you do not trust the surgeon, or you do not want any cuts or scars on your skin. All of these have nothing to do with probabilities.

The following is a true story from World War II (WWII) regarding a decision between two seemingly equally likely alternatives.

1.3.1 *How would you choose between two equal-probability alternatives?*

The first time I heard this story was in a lecture on game theory that was given by the late Michael Maschler, a mathematics professor at the Hebrew University in Jerusalem. At that time, Maschler cited Kenneth Arrow as the source of the story. While preparing this book, I wrote to Arrow in order to verify the details, but all he could say was that he had heard the story from Merrill Flood, who was an Operations Research Officer with the United States Air Force in the Pacific during WWII. More recently, I found this story quoted in an article by Robert

Fig. 1.9 Pilots in WWII.

Aumann titled *Rule-Rationality versus Act-Rationality*. Although I have not been able to validate the exact details of the story, the pilots' dilemma and the way the probabilities were subjectively perceived are correct.

This is a true story about a squadron of American bombers who were assigned to the Pacific Islands during WWII whose mission was to carry out bombing sorties on an island that belonged to Japan and was about 800 miles away. Owing to this great distance, much of the weight that the bombers could carry was given over fuel, and a lesser weight was for bombs. For concreteness, let us assume the following.

A fighter plane's maximum load capacity was, at that time, say, 3 tons. Of this 3-ton capacity, 2 tons of fuel was needed for a round trip, which leaves only 1 ton for bombs. Statistically, it was known that 50% of the planes would be shot down, killing the pilots. Thus, all of the pilots knew that when they took on these missions, they had a 50% survival rate.

With the aim of enhancing the impact and efficiency of the bombing sorties on the target islands, the Operations Research

Officer offered the pilots two choices: The first was to go for a regular mission with a full load of fuel, i.e. 2 tons of fuel and 1 ton of bombs. The second choice was to toss a fair coin. If the result was a "Head," the pilot would be "off the hook," meaning he was free to go home and not participate in the mission. A "Tail" would mean that the pilot was to go ahead with the mission but with only 1 ton of fuel, which was only good for a one-way trip, and 2 tons of bombs. Clearly, the two choices are statistically equivalent. In each of these cases, the pilots had a 50% chance of survival, and yet the pilots unanimously chose the first option.

What would you have chosen?

Kenneth Arrow told me that, as far as he could remember, the odds were actually very much in favor of the one-way mission. Statistically, the fraction of pilots who took the mission and returned was only 1/4. Therefore, by accepting the Operations Research Officer's offer, the pilots would have *increased* their chances of survival from 1/4 to 1/2.

Nevertheless, the pilots refused the offer of the Operations Research Officer. This behavior seems to be irrational. Why would one choose an option giving lower chances of survival? When asked during the individual interviews as to why they refused the offer, each one replied that *he felt that he was a much better pilot than the average*, and therefore *he* would not be shot down.

Biased assessments of the chances of survival play out quite often in our everyday lives. Every time someone drives a car at a high speed, the chances of a car crash resulting in death are high and yet we are somehow "programmed" to assess that these statistics apply to everyone else and not to us.

Long ago, I read in a preface of a book about the theory of traffic flow. The theory was quite useful in designing bridges and highways. It assumes that people drive at a random speed, making turns at random times and at random places. The author commented that it is very easy to convince the reader that *all* drivers behave randomly, but it is impossible to convince an individual that he or she is driving at random speed, making random turns at a random time and place.

1.4 Perception of Rare Events

Our personal subjective assessments are understandable. Furthermore, the subjective feelings of *all* of the pilots that they were *above* the average is understandable. However, there is another aspect of probability that is less understood: How one perceives the occurrence of rare events or rare coincidences of events. We are all intrigued by coincidence. In ancient times, extraordinary coincidences were perceived as miracles or the actions of some supernatural forces.

If you are told that someone in a remote village won the 10 million dollar prize in a lottery, you would not be impressed because, after all, you know that someone out of the millions who bought tickets to the lottery will likely win. However, when you are told that *you yourself* won the jackpot, you are stunned because you cannot help but believe that some divine power has intervened in your favor.

When I was a boy, I had a friend who also happened to be our next-door neighbor. She once told me that she was endowed with some extraordinary mental capabilities. She narrated to me how she dreamt that her grandmother had died, only to be frantically roused by her mother the morning after, informing her that her grandmother had died the night before.

I was not impressed by her story, nor by her claim of possessing extraordinary mental powers. Somehow, I sensed that many of us dream about someone dying in the family, wake up and forget all the about the dream. There are of course many stories about dreams that actually happen in real life. This does not mean however that some supernatural power is involved.

Here is another story that I took from Ruma Falk's article (1981–1982):

The Nobel laureate physicist Luis Alvarez (1965) related, in a letter to *Science*, a strange event that happened to him: A casual phrase in a newspaper triggered a chain of personal associations, that reminded him of a long-forgotten figure from his youth. Less than five minutes later, having turned the pages of the paper, he came across the obituary for that same person. Alvarez noted that these two closely spaced recollections of a person forgotten for 30 years, with the second event involving a death notice, follow the classical pattern of many popular parapsychology stories. Such coincidences often make people feel that there must be a causal relation between the two events, as, for example, by thought transference. However, considering the details of this case, it was obvious that no causal relationship could have existed between the two events.

Alvarez felt that scientists should bring such stories to public awareness to show that such apparently improbable coincidences do, in fact, occur by chance. He also endeavored to compute its probability by estimating the number of people that an average person knows and the average frequency with which one remembers one's acquaintances. Using rather conservative assumptions, he arrived at the conclusion that the probability of a coincidental recollection of a known person in a 5-minute period just before learning of that person's death is 3×10^{-5} per year. Given the population of the United States, we could expect 3000 experiences of the sort to occur every year, or about 10 per day. Considering coincidences of other kinds as well increases this number greatly. "With such a large sample to draw from, it is not surprising that some exceedingly astonishing coincidences are reported in the parapsychological literature as proof of extrasensory perception in one form or another" (Alvarez, 1965). Paradoxically, the world is so big that quite a few of these "small-world-phenomena are bound to occur."

1.5 My Father's Perception of "Probability"

I would like to end this introductory session with two stories about my father. One involves perceptions of probabilities as signs from some divine power. The second involves the 'divine power' of garlic.

My father used to buy one lottery ticket every weekend for almost 60 years. He was sure that someone "up there" favored him and that someday he would be "rewarded" with the grand prize. I repeatedly tried to explain to him that his chances of winning the grand prize were very slim, in fact, less than one hundredth of one percent. But all my attempts to explain his odds to him fell on deaf ears. Sometimes he would get seven or eight matching numbers (out of ten, with ten matches being the winning combination). He would deride me for not being able to see the clear and unequivocal "signs" he was receiving from Him. He was sure he was on the right track to winning. From week to week, his hopes would wax and wane according to the number of matches he got, or better yet, according to the kind of signs he believed he was receiving from up above. Close to his death, at the age of 96, he told me that he was very much disappointed and bitter, as he felt betrayed and disfavored by the deity in whom he had believed all his life. I was saddened to realize that he did not, and perhaps could not, think *probabilistically*!

The second story involves my army days as a young soldier. As was and still is practice in the army, soldiers were allowed days off, which I always spent with my family. As soon as I was ready to go back to the base, my father would ask me to put a clove of garlic in my pocket, saying that it would keep me safe. The practice ensued, even after I had gotten married and left home. There were even times when he surreptitiously dropped a clove of garlic into my pocket.

At that time, I thought he believed that garlic had magic powers to ward off the angel of death, or perhaps minimize the chances of the angel of death getting close to me. Could that have been probabilistic thinking? While writing this book, I pondered about this topic again and came to the conclusion that there was nothing probabilistic about it. My father was *certain* (probability of 1!) that garlic was effective in keeping the angel of death at bay! (Could it be because of its repulsive smell?)

1.6 Conclusion

In this chapter, we saw that each of us is endowed with some kind of rudimentary sense of probability that guides us in making a decision between two or more choices. The concept of "probability" (or "randomness," "chances," "odds," etc.) has its roots in very ancient times. An interesting and fascinating account of the history of the concept of probability may be found in David (1962) and Bennett (1998).

We also learned that the meaning attributed to chance events has evolved from being seen as controlled by a divine power, to being seen as purely random, as we conceive of them today.

Session 2

How Do We Calculate Probabilities?

I hope that you are now comfortable with the intuitive meaning of probability. I also hope that you gave the correct answers to all of the questions I asked you in the previous session. The answers you gave were based on intuition, or on your sense of probability. This sense guides you in finding the correct probabilities for these very simple cases. In more complicated examples, intuition alone is not sufficient. You will need some mathematics for these more complex cases. For now, we shall deal with only simple cases in which the calculation of probabilities is quite straightforward.

2.1 The Classical "Definition" of Probability

Note that I have enclosed "definition" in inverted commas. You will understand why later on. For the moment, let us "invent" or "discover" this definition by ourselves.

Example: We throw a fair dice in such a way that each outcome has the same likelihood of occurrence. What are the probabilities of the following events? (Note that here we use curly brackets for the event "4": {4}. This is consistent with the notation in set theory. Later, we shall use either {4} or "4" to denote that the result "4" occurred.)

(a) The outcome is {4}.
(b) The outcome is an even number, i.e. it is one of the results {2} or {4} or {6}. We write this event as {2, 4, 6}.
(c) The outcome is greater than 4. This means it is either {5} or {6}. We write this event as {5, 6}.

Write down your answer before you check Note 1.

How did we assign probabilities to these events? We assumed that the dice is "fair" and that we threw it in such a way that each single possible outcome had the same likelihood. This means that the dice is a perfect cube, its mass density is evenly distributed and that we threw the dice in such a way that it rolls and spins in the air many times so that it will eventually land on the floor with one of its sides facing upwards.

Before we calculate the probability of an event, we have to decide on the *range* of values that the probabilities can attain. The choice of the range is arbitrary. This is similar to the choice of scale for temperature. The most common choice of range is between zero and one (see also next session). In our daily lives, we often use the scale between 0% and 100%. The lowest probability value zero is assigned to an *impossible* event. The highest probability value one is assigned to a *certain* event.

Example: In throwing a regular dice, the outcome "17" has probability *zero* of occurrence. The outcome {1, 2, 3, 4, 5, 6} — i.e.

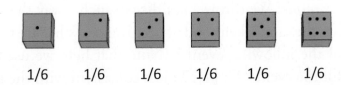

Probabilities: 1/6   1/6   1/6   1/6   1/6   1/6

Fig. 2.1   All possible outcomes of a dice and their probabilities.

either "1," "2," "3," "4," "5" or "6" — has probability *one*. We of course assume that these are the only possible outcomes, and ignore the possibility that the dice will fall on one of its edges or on one of its vertices.

Now for event (a) in the example above, we reason that since there are altogether six possible outcomes, and we assume that each outcome has the same likelihood of occurrence, the probability of a single outcome, say {4}, in case (a) is 1/6.

If you pause and think about the reasoning that led us to the number 1/6, you will find that we used the phrase "each of the outcomes has the same "likelihood"," which is tantamount to saying that each event has the same probability. Once we have also fixed the value of the *certain* event to be one, we can calculate the probability of each single event as being 1/6. Thus, what we have done is not to *define* the probability of the event {4}, but to assume that we *know* the probability of the event to be 1/6. In other words, this "definition" is *circular*. It uses the concept of probability (or likelihood, chances or odds) to "define" the probability. Sometimes an argument based on symmetry or equivalence of all possible results is used to reach the conclusion that the probability of each event is 1/6.

Let us go to case (b). What is the probability of the occurrence of the event {2, 4, 6}, i.e. that the outcome is an *even* number?

There are two ways of reasoning about this. First, we can argue that there are altogether six equally likely outcomes, each having probability of occurrence 1/6. Therefore, the occurrence of either {2} or {4} or {6} must be larger than 1/6, and it is most likely to be the sum of these three probabilities, i.e. $\frac{1}{6} + \frac{1}{6} + \frac{1}{6} = \frac{3}{6} = \frac{1}{2}$.

The second way of reasoning is to divide all of the possible outcomes into two groups of events: "even" and "odd" outcomes. Think of coloring all of the faces of the dice with even numbers

Fig. 2.2 A dice having three red and three blue faces.

Fig. 2.3 A dice having two red, two blue and two green faces.

of dots in red, and all the faces of the dice with an odd number of dots in blue. The probability of the event "even" is equivalent to the probability of the event "red" face. Since there are two possible outcomes — either "red" or "blue" — and since we believe that the dice is fair, we conclude that the probability of the event "red" (or "even") is 1/2.

Note again that in calculating the probability of the "event" we used *probabilistic* arguments, i.e. we assumed that each outcome had the same likelihood. Therefore, this method of calculation cannot be viewed as a bona-fide definition of the probability of the event "even."

Let us turn to case (c). The event "greater than" 4 means that the outcome is either {5} or {6}. We write this event as {5, 6}. Using the same type of argument as before, we can conclude that the probability of the event {5, 6} is the sum of the probabilities of the single outcomes {5} and {6}, i.e. $\frac{1}{6} + \frac{1}{6} + \frac{2}{6} = \frac{1}{3}$.

Exercise: Suppose that the faces of the dice are colored as follows: Faces {1} and {2} in red; faces {3} and {4} in blue; and faces {5} and {6} in green. What is the probability of the event "green"?[2]

If you have calculated the probabilities of the events in (a), (b) and (c) in the example above correctly, you have *almost* discovered the so-called *classical definition* of probability.

Before we continue with the formal "definition" of probability, let me tell you about two experiments conducted on young children in order to test their understanding of probability.[3]

Children in grades 5, 7, 9 and 11 were told the following: In a lotto game, one has to choose six numbers from a total of 40 numbers, from 1 to 40. Vered chose six numbers: $\{1, 2, 3, 4, 5, 6\}$, while Ruth chose the following six numbers: $\{3, 9, 1, 17, 33, 8\}$.

Who has a greater chance of winning?

The following answers were suggested:

(a) Vered has a greater chance of winning.
(b) Ruth has a greater chance of winning.
(c) Vered and Ruth have the same chance of winning.

The percentages of students choosing the different answers are given in Table 2.1 below

Table 2.1

Answer	Percentages of Student Answers				College Students
	Grades				
	5	7	9	11	
(a)	0	0	0	0	0
(b)	70	55	35	35	22
(c)	30	45	65	65	78

These answers show the common misconception of *representativeness*. The students tend to estimate the likelihood of an event by taking into account how well it *represents* some aspects of its parent population. Clearly, this misconception was large among the younger students and decreased with age. Of course, the correct answer is (c). Note that the percentage of the students answering (c) increases with age.

Take a look again at the percentages in the table above. See how younger children believe (or sense) that the "ordered" sequence of number is more unlikely to occur compared with the random sequence, or the more "representative" sequence of numbers.

Another experiment was designed to test the intuitive understanding of the probability of *compound events*. The question is the following: We throw two fair dice simultaneously. Which of the following has the greater chance of happening?

(a) Getting the pair 5 *and* 6.
(b) Getting the pair 6 *and* 6.
(c) Both have the same chance.

The following answers were obtained:

Table 2.2

Answer	Percentages of Student Answers				College Students
	Grades				
	5	7	9	11	
(a)	15	20	10	25	6
(b)	0	0	0	0	0
(c)	70	70	75	75	78

No one gave the incorrect answer (b). The correct answer is (a). The probability of obtaining the pair 5 and 6 is twice than that of the pair 6 and 6. (Remember that the order of the outcomes does not matter. Make sure you understand why answer (a) is the correct one.) Yet most students gave the incorrect answer (c), and it seems that this misconception is stable, because in the general population, this percentage of responses does not change with age. We shall further discuss these kinds of questions in Session 3.

I urge you to pause and take a look again at the figures in Table 2.2 above, and draw your own conclusion based on these data.

The classical "definition" of probability is: For an experiment that has n equally likely outcomes, denoted $A_1, A_2, \ldots, A_n$, the probability of an *event* B is calculated by the rule:

$$Pr(B) = \frac{Number\ of\ outcomes\ included\ in\ B}{Total\ number\ of\ outcomes}.$$

Each single possible outcome is said to be an elementary event.

Let us first see that our calculations for the events in (a), (b) and (c) in the example given on page 28 are consistent with this rule.

In case (a), the event "B" consists of one outcome: {4}. Applying the rule above, we get:

$$Pr(a) = \frac{1}{6}.$$

In the case (b), the event "B" contains three outcomes, hence:

$$Pr(b) = \frac{3}{6} = \frac{1}{2}.$$

In case (c), the event "B" contains two outcomes, hence:

$$Pr(c) = \frac{2}{6} = \frac{1}{3}.$$

We see that the classical "definition" is intuitively clear. You should realize, however, that this is not a *bona-fide definition* of probability. The "classical definition" already assumes that we *know the probabilities* of each single outcome. Therefore, this definition is circular.

Furthermore, this rule of calculating probabilities does not apply in general. First, it is not always clear what the *elementary outcomes* are. We shall see examples of such cases in the next session. For now, it is sufficient to say that, in the case of throwing a dice, we assume that there are six possible outcomes (we neglect the possibilities that the dice will fall on an edge or on a vertex, or that perhaps it will fall and break into pieces so that no clear outcome is observed).

More importantly, elementary events are not always equally probable. We shall further discuss these cases in the next session. Therefore, the classical definition — or rather, the method of calculating probabilities by the rule given above — does not apply for all cases. Nevertheless, it is a very useful rule for calculating the probabilities of a large class of cases. We have seen the relatively simple example for a dice. There are more complicated cases in which this rule is useful. Sometimes, the calculation of the numbers in the numerator and the denominator of the *rule* is extremely difficult. The branch of mathematics that deals with such problems is called *combinatorics*. It consists of mathematical methods of calculating the number of ways of selecting or arranging a group of n objects out of N objects with some constraints or some restrictions.

Let us conclude this section with two slightly more difficult problems.

Exercise: An urn contains six red, four blue and three yellow marbles (see Fig. 2.4). You know the contents of the urn. You close your eyes and shake the urn (so you will not be able to use the knowledge you might have on the locations of the marbles in

Fig. 2.4 An urn with six red, four blue and three yellow marbles.

the urn). You draw two marbles at once. What is the probability that, when you open your hand, you will find one red and one blue marble?

In order to calculate the probability of the required event, we use the rule as written above.

Before doing the calculation, we must first determine what the "elementary outcomes" are, i.e. those outcomes that are equally probable. Consider all possible pairs of colors. These are:

{red, red},   {blue, blue},   {yellow, yellow},   {red, blue},

{red, yellow},   {blue, yellow}.

Clearly, these events are not equally probable. These are indeed all possible *pairs of colors*, but the pair {red, red} is more likely to occur than the pair {yellow, yellow}. Therefore, this is not the right choice of the elementary outcomes. Instead, we can calculate the number of all *elementary outcomes* as follows. Assuming that the marbles have the same size and feel the same when you pick any one while you are blindfolded, we can reasonably assume that each pair of two *specific* marbles has the same probability of being drawn. Now the total number of *specific* pairs can be calculated as follows:

$$\text{Number of \{red, red\}} = 6 \times \frac{5}{2} = 15$$

$$\text{Number of \{blue, blue\}} = 4 \times \frac{3}{2} = 6$$

$$\text{Number of \{yellow, yellow\}} = 3 \times \frac{2}{2} = 3$$

$$\text{Number of \{red, blue\}} = 6 \times 4 = 24$$

$$\text{Number of \{red, yellow\}} = 6 \times 3 = 18$$

$$\text{Number of \{blue, yellow\}} = 4 \times 3 = 12.$$

Thus, for the {red, red} pair, there are six *possibilities* of drawing the first red marble, and for each of these draws, there are *five* possibilities of drawing the second red marble. Since the *order* of the marbles is not important, we have to divide the product of 6 × 5 by 2 in order to obtain the total number of pairs of {red, red}. Convince yourself that this calculation is correct. Try different numbers of marbles of each color. Note also that these calculations belong to *combinatorics*; these are exact numbers. We construct the relevant probabilities after we calculate the *total* number of pairs, then use the classical definition of probability.

The total number of equally probable outcomes is the sum of all the above numbers, i.e. 78. Out of these 78 outcomes, there are 15 outcomes that are {red, blue}. Therefore, according to the "classical definition," the probability of the required event {red, blue} is:

$$Pr(\text{red, blue}) = \frac{24}{78}.$$

If you like to think in more abstract terms, suppose there are NR red marbles, NB blue marbles and NY yellow marbles. The

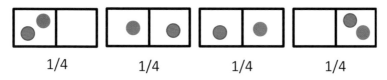

    1/4   1/4   1/4   1/4

Fig. 2.5 All possible configurations for two different marbles in two compartments.

list of the total number of marbles of each pair of colors will now be provided in Note 4.

The next example is important to many problems in physics. It is related to one of the most bizarre behaviors of particles in the microscopic world. We shall discuss here the simplest example in order to demonstrate that sometimes it is difficult to decide what the elementary outcomes are. Suppose we have a box with two compartments (see Fig. 2.5). The two compartments have the same size and shape, such that if we place a marble in the box and shake the box vigorously in all possible directions, the marble is *equally likely* to be found in either the right or the left compartment, which in this case is a probability of 1/2.

Now we place two different marbles, say one red and one blue, in the box. The marbles have the same size, shape, weight, etc. We assume that the marbles are much smaller than the box so that the presence of one marble in a compartment does not affect the probability of the other marble being in the same compartment. Again, we assume that if we shake the box vigorously, there is probability 1/2 of finding the red marble in either compartment and 1/2 of finding the blue marble in either compartment. See Fig. 2.5 for all possible cases.

After shaking the box so that the marbles are well "mixed," we take snapshots of the contents of the box. We are not interested in the exact point in space in which each marble is located, but only in which of the two compartments each marble is found. There are four possible outcomes, as shown in Fig. 2.5.

What is the probability of finding each of these outcomes?

It is reasonable to assume that these four possible events are equally probable. Therefore, the answer to the question is immediate: Each of the outcomes in Fig. 2.5 has probability 1/4 (one out of four snapshots).

Now, let us look at the trickier question. Suppose that the two marbles are *identical*: Both have the same color, shape, size, weight, etc. What are the elementary, equally probable events? Note that in this case we can *distinguish* between three different *configurations*. We can obtain these configurations simply by erasing the colors of the marbles in Fig. 2.5. The second and the third outcomes in Fig. 2.5 coalesce into one (see Fig. 2.6), i.e. in this case, there are only three distinguishable outcomes, as shown in the second row of Fig. 2.6. These are: Two in the left compartment, two in the right compartment and one in each compartment (the two configurations where one marble is in each compartment are indistinguishable).

What are the probabilities of these outcomes?

If the marbles are macroscopic objects, like real marbles, then although the two marbles are *identical*, the three distinguishable

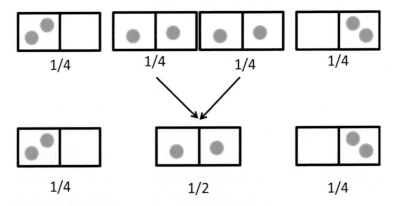

Fig. 2.6 The four different configurations from Fig. 2.5 coalesce into three when the marbles are identical.

configurations in Fig. 2.6 are not equally probable. Each marble has equal probability of being either in the right or in the left compartment. Therefore, if we shake the box and take many snapshots, we shall find the two marbles in the left compartment in one out of four snapshots, i.e. with probability 1/4. The same is true for finding the two marbles in the right compartment. However, the probability of finding one marble on the right and one on the left compartment is 1/2. Thus, the three distinguishable configurations shown in the lower panel of Fig. 2.6 are not equally probable.

We can summarize what we have found so far as follows:

The following *three* events (or configurations) are:

(1) Two marbles in the right compartment.
(2) Two marbles in the left compartment.
(3) One marble in the left and one in the right compartment.

These events have *different* probabilities of occurrence, and therefore cannot be viewed as elementary events. However, the following *four* events (or configurations) are equally likely to occur:

(a) Marble labeled (1) in the right and marble labeled (2) in the right;
(b) Marble labeled (1) in the right and marble labeled (2) in the left;
(c) Marble labeled (1) in the left and marble labeled (2) in the left;
(d) Marble labeled (1) in the left and marble labeled (2) in the right.

Each of these four events has the same probability 1/4, independently of whether the marbles have the same or different colors and whether they are labeled by numbers or not.

We started our reasoning in this section by saying that the marbles are *macroscopic* objects. In this case, although the marbles are *identical*, they are said to be *distinguishable* in the sense that while we shake the box, we can in principle follow each of the marbles as if they were *labeled*. We can, for instance, start with any configuration, e.g. with each marble in a different compartment. While we shake the box, we can in principle *follow the trajectory* of each of the marbles. Thus, the marble that was initially in the left compartment, after being shaken, has equal probability of being found either in the right or the left compartment. The same is true for the marble that was initially in the right compartment.

If the marbles are *microscopic* — of the size of atoms or electrons — then a strange phenomenon occurs. In the early 20th century, it was discovered that microscopic particles behave differently from macroscopic particles. New theory has been developed to deal with molecular phenomena. This new theory is called quantum mechanics. For the particular example we are discussing, the principle of quantum mechanics states that we cannot follow the trajectory of each particle separately. We say that the two particles are *indistinguishable*. This means that the two configurations (b) and (d) are indistinguishable, i.e. the two particles cannot *in principle* be labeled. In the microscopic world, the four configurations listed above are not elementary events; they do not have equal probabilities.

This principle was discovered "experimentally." It was found that calculations based on the assumption that all four configurations in Fig. 2.6 are equally probable, yielded results which did not agree with experimental results. The culprit for these wrong results was traced back to the assumption of equal probabilities

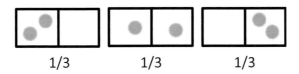

Fig. 2.7 Three configurations having equal probabilities.

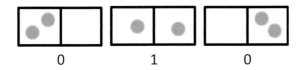

Fig. 2.8 Three configurations having un-equal probabilities.

to all four events (a), (b), (c) and (d). What was found was far more bizarre than expected. Not only are the four events (a), (b), (c) and (d) not equally probable, but the probabilities of these four events depend on the *type* of the particles. In the microscopic world, the "compartments" are replaced by quantum mechanical *states*. For simplicity, we shall continue to use the language of particles in different "compartments." It was found that there are two kinds of particles obeying different laws. The first kind are called bosons, and are said to obey Bose–Einstein statistics. In this case, the three configurations in Fig. 2.7 have equal probability of 1/3. The second kind of particles are called fermions, and are said to obey Fermi–Dirac statistics. In this case, there is only one elementary event, which is the configuration shown in Fig. 2.8. The two particles are not allowed to occupy the same compartment (in the microscopic world, the "compartment" is a quantum mechanical state, or an *orbital*, and two fermions are not allowed to occupy the same state). It is as if the two fermions repel each other and cannot "live" in peace in the same compartment.

The fact that two electrons, which are fermions, cannot occupy the same quantum mechanical state ('compartment') is called the Pauli Exclusion Principle. This principle has far reaching consequences in physics and chemistry. The very existence of the various elements in our universe depends on this principle. Without this principle, the periodic table of elements would not exist. There would be no hydrogen, oxygen, nitrogen or carbon atoms. There would be no water, proteins or DNA, and of course there would be no life as we know it on our planet. Someone said that without the Pauli principle, our world would "look very different." In fact, without this principle, there might not be anyone to "look" at this "world."

Although the study of the behavior of bosons and fermions is not part of the theory of probability, I urge you to do the following simple exercise:

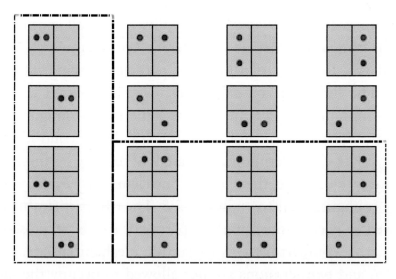

Fig. 2.9 All possible configurations.

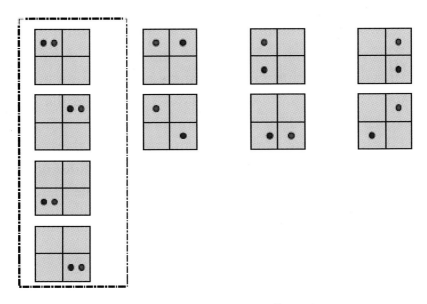

Fig. 2.10   Bose-Einstein configurations.

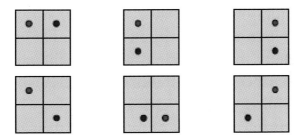

Fig. 2.11   Fermi-Dirac configurations.

You have two particles and four compartments as shown in Fig. 2.9. List all of the possible configurations and the corresponding probabilities for the case of the two marbles in boxes, and for two fermions and two bosons in four quantum mechanical "boxes" (see also Figs. 2.10 and 2.11).[5]

2.2 The Relative Frequency "Definition"

In all of the examples discussed in the previous section, we started with the assumption that there exists a finite number of elementary events, or elementary outcomes, and that these have equal likelihood (or chances or probability) of occurrence. How do we calculate probabilities in cases where there are no obvious elementary events? How are the probabilities of such events defined? The general and honest answer is that there is no *bona-fide* definition of probability, nor a method of calculation that is satisfactory for general events.

What is the probability that a dormant volcano will erupt in the next hour? What is the probability that the sun will explode tomorrow? What is the probability that a child will be born with four legs, six arms and three heads?

Clearly, there is no way of defining, let alone calculating, the probabilities of these events. Yet people do use the term "probability" in connection with such events. The only meaning that "probability" has, in such a context, is the extent of one's belief in the chances of occurrence of these events.

However, there is a large class of events for which one can offer an "experimental" way of calculating their probabilities. These are the cases in which we can repeat an experiment many times, or certain events have occurred many times in the past, and we can collect "statistics" on specific events. For example, suppose we have a dice that is known to be unfair, say with an asymmetric distribution of mass, or a partially broken or twisted dice. Obviously, we cannot assume that each outcome has the same probability in this case.

In this particular example, we apply the so-called *relative frequency* "definition" of probability.[6]

We throw the dice many, many times (say a thousand times) and collect the "statistics" about the frequency of observing the results {1}, {2}, {3}, . . . , {6}. By "frequency," we mean the ratio of the number of times a specific result occurrs, and the total number of throws.

Suppose we found the following results after a thousand throws:

$$50 \text{ results showing } \{1\}$$
$$100 \text{ results showing } \{2\}$$
$$100 \text{ results showing } \{3\}$$
$$200 \text{ results showing } \{4\}$$
$$250 \text{ results showing } \{5\}$$
$$300 \text{ results showing } \{6\}.$$

We might tentatively *assume* that the probabilities of the different outcomes are:

$$\frac{50}{1000}, \quad \frac{100}{1000}, \quad \frac{100}{1000}, \quad \frac{200}{1000}, \quad \frac{250}{1000}, \quad \frac{300}{1000}.$$

The relative frequency definition states that if we throw the dice *infinite* times, the fraction of times each outcome occurs is the probability of that event.

This definition is problematic at best. In fact, it is also circular in the following sense. We believe that if we do the experiment infinite times, then the fraction of times each outcome occurs will tend to some constant value between zero and one. Unfortunately, we cannot perform an infinite number of trials or experiments. In fact, no one can guarantee that if we calculate these fractions for many experiments, the fractions will tend to some constant values.

In practice, we make a large but finite number of experiments, collect the "statistics," as we did above, and *assume* that these results are the approximate probabilities of the outcomes. We believe that if we repeat the experiment many times (thousands, millions, billions . . .), it is *highly probable* that the fractions we would get would be the "true" probabilities. You see that we use the concept of "probable" to define the concept of probability. Therefore, this method cannot be considered to be a *bona-fide* definition.

However, in practice, we use this method with finite numbers of experiments in order to estimate the most likely probabilities. It is not perfect, it does not guarantee that we will get the "correct" results and it is not always applicable. Yet *this is what we have*, and in many cases, this method is very useful.

Based on this method, we can determine the approximate probabilities of outcomes of an unfair dice, as we did before. Doctors and pharmaceutical companies determine the efficacy of certain drugs. Insurance companies estimate the likelihood that a certain person (within a specific age bracket, sex, education level, marital status, etc.) will be involved in an accident, and with this estimate they calculate the cost of one's insurance policy. In all of these cases, and in many others, we do not have "exact" probabilities, but this is the best we have, and we use these likelihoods because we *need* to use them.

Let us do some exercises. Suppose I throw an unfair dice a thousand times and found the following results:

The outcome {1} occurred 600 times.

The outcome {2} occurred 98 times.

The outcome {3} occurred 96 times.

The outcome {4} occurred 95 times.

The outcome {5} occurred 97 times.

The outcome {6} occurred 14 times.

Based on the gathering of these statistics, estimate the probabilities of each of the outcomes. What is the probability of the event "even"? What is the probability of the event "odd"? Can you guess how I prepared this dice?

Based on the given data, my estimate of the probabilities are:

Event {1}: 600/1000

Event {2}: 98/1000

Event {3}: 96/1000

Event {4}: 95/1000

Event {5}: 97/1000

Event {6}: 14/1000.

The probability of the event "even" is:

$$\frac{98 + 95 + 14}{1000} = \frac{207}{1000}.$$

The probability of the event "odd" is:

$$\frac{600 + 96 + 97}{1000} = \frac{793}{1000}.$$

You may guess that I took a regular dice and added a heavy metal to the face with six dots so that this face will, with high probability, land facing the floor, while the face "1" will land facing upwards. In this case, the probability of the outcome {1} will be the largest, the probability of the outcome {6} will be the smallest and all other faces have nearly equal probabilities of about 1/10.

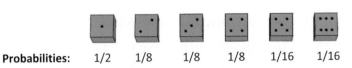

Probabilities: 1/2   1/8   1/8   1/8   1/16   1/16

Fig. 2.12   An unfair dice.

Note that this method only gives a reasonable estimate of the probabilities. It does not guarantee that these are the "correct" probabilities. We believe that had we done one million throws or one billion throws, we would get results that are closer and closer to the "correct" probabilities. But the truly correct probabilities are elusive. Even if we do infinite throws (whatever that means), we cannot be assured that we will get the "correct" probabilities (whatever that means as well). Yet, in spite of this uncertainty, this is the most useful and most used method of calculating probabilities. It is useful not because it is accurate, but because this is the *best* method available to us.

You might wonder how this method compares with the classical "definition." The latter sounds more accurate, more precise and more reliable, but this is only an illusion.

First, we can never be sure that the dice is perfectly fair (whatever that means). If we are not sure, we must use the experimental method, and if we find that each outcome occurs with the same frequency, i.e. about $\frac{100}{600}$, or $\frac{1000}{6000}$, etc, then we can be reasonably sure that it is a fair dice, and that the probabilities are 1/6. But what if we are (somehow) sure that the dice is fair, how do we know that the probabilities of the outcome are equal to 1/6? In fact, we do not *know*. We believe that this is a reasonable assumption. If we doubt this assumption, we can do the experiment again, or we can imagine running the experiment many times. Of course, in our *imagination* we can afford to repeat the same experiment an *infinite* number of times, and *imagine* that the relative frequencies will all be equal to 1/6.

Let us do another mental exercise.

Suppose a doctor has two drugs to prescribe for a certain disease. The physician obtained the information from the drug companies.

Drug A was administered to 100 people and it was found that the drug was effective in 60 out of 100 subjects, and that these patients recovered totally from the disease.

Drug B was administered to a million people who were afflicted with the same disease, and it was found that 60% of those who took the drug recovered from the disease.

Which drug would you choose if you were the doctor?[7]

Finally, consider the following exercise (we shall learn how to "modify" the estimated probabilities based on new information in Session 4. But for now, just read and consider what you would have chosen if you were the patient).

You have the following information provided by the drug company:

Drug C was administered to 1000 patients and was found effective for 800 of them.

Drug D was administered to 1000 patients and was found effective for 700 of them.

Which drug would you choose?

It seems that drug C is better than drug D for this specific disease. Therefore, it is reasonable to prefer drug C.

Now you decide to do some research. You search the internet and read the literature, and you find more details on how the drug company carried out their experiments.

Drug C was administered to 500 men and 500 women; 300 of the men recovered and 500 of the women recovered.

Drug D was administered to 500 men and 500 women; 500 of the men recovered and 200 of the women recovered.

"Which drug will you choose if you were a man?"
"Which drug will you choose if you were a woman?[8]"

Exercise: To think about, but not to do.

This problem does not really belong in this session. It is an easy problem if you know about independence between events (Session 4). It is more difficult if you have never heard about independence between events.

Do not try to solve this problem late at night, as it might keep you awake.

You are given an urn with 10 marbles. The marbles are of the same size but are numbered from 1 to 10. While your eyes are closed, you draw a single marble, look at its number and write it down, then return the marble to the urn. You then shake the urn, draw another marble, write its number, and so on, until this process had been done ten times. Calculate the probabilities of obtaining the following sequences of numbers in ten draws:

(a) 1, 2, 3, 4, 5, 6, 7, 8, 9, 10
(b) 4, 4, 4, 4, 4, 4, 4, 4, 4, 4
(c) 3, 7, 9, 2, 4, 7, 1, 5, 6, 2.

Note that the order of the outcomes is important. I am asking about the probability of a *specific* order of numbers in each case (hint: Before calculating the actual probabilities, try to estimate the relative probabilities of each sequence).[9]

2.3 Conclusion

In Session 1, we learned that there is no definition of the term "probability." However, we also saw that we have some kind of a sense of probability that guides us in making some probabilistic decisions. In this session, we learned that there is no absolutely reliable method of calculating probabilities. Instead, we rely on or believe that if we repeat an experiment many times, the resulting frequencies will be a reasonable measure of the extent of certainty or uncertainty regarding the occurrence of one outcome or another. It is clear that we live in a world of uncertainty. This session taught us that we are also uncertain about the extent of our uncertainty of the occurrence of an event.

Yet in spite of all these uncertainties, shortcomings, limitations and so on, we need probabilities in almost every aspect of our lives. Without knowledge of probability theory, we might not be able to make prudent decisions, or we might be led into making wrong decisions. Probability theory offers us the best it can, and in many cases it is extremely useful. Without probabilities, a doctor would not be able to decide on which drug to prescribe to their patients. Insurance companies would not be able to estimate the cost of insuring cars, houses or life insurance. Food companies rely on feedback and statistics in order to improve or change their products, and in some cases to come up with new ones. Traffic lights are designed to switch from green to red at different intersections, at different times of the day based on collected statistics on traffic flow at each intersection. It is difficult to find even one aspect of our lives that does not depend on probabilistic reasoning and decision making. Look at the illustrations at the end of this session. For each of these, try to "invent" a relevant probabilistic question.

It is often said that there are two certain things in life: Taxes and death. I think a better way of saying this is that taxes and death are certain only with *high probability*.

We shall see in subsequent sessions that acquiring some proficiency in probability could be crucial in making decisions regarding life or death situations. More precisely, the decision could increase or decrease our chances of survival or death.

There are different alternatives in the following drawings. Suggest a probabilistic question relevant for each.

Session 3

The Axiomatic Approach to Probability

As in any other branch of mathematics, one starts with a few axioms. These are essentially postulates upon which everyone agrees. One builds up an entire theory based on these postulates. It should be noted that probability theory is relatively a "newborn" theory in mathematics. In was only in the 1930s that the axiomatic theory of probability was established. In this session, we present the so-called axiomatic approach to probability. Some authors refer to this approach as a *definition* of probability. It is not! This approach does not *define* probability, but only assumes that there is a quantity which we call probability that satisfies some requirements, as we shall see below.

3.1 The Probability Space: (Ω, L, P)

The probability space is not a *probability* nor a *space*. But do not worry, by the time you reach the end of this section, you will know what it is, and what it contains. As the title of this section implies, the probability space, which is not yet defined, contains three "elements" or "ingredients," denoted by the symbols Ω, L and P. These three "symbols" are just *short-hand symbols*, and do not imply any mathematical content. Let us see what each of these symbols means.

3.2 The Sample Space: Ω

Consider an experiment or a game, the outcomes of which are denoted by the set $\Omega = \{A_1, A_2, \ldots, A_n\}$.

Examples include:

(i) Tossing a coin has two outcomes, heads (H) and tails (T). We ignore any other possible (though very rare) outcome, such as the coin falling exactly on its side, breaking into pieces or completely disappearing. Thus, we assume that there are only two possible outcomes, and we denote the set of outcomes by $\Omega = \{H, T\}$.

Throwing a dice has six outcomes. We assume that when the dice lands on the floor, it shows one and only one of its faces. The faces might have different numbers of dots or might have different colors. In any case, we denote the set of the six outcomes by $\Omega = \{1, 2, 3, 4, 5, 6\}$. Again, we assume that when we throw the dice, one and only one of these outcomes occurs. We ignore many other possibilities (that the dice will land on an edge, on a vertex or will disintegrate into small pieces, etc.).

(ii) Note that the faces might have different colors, or have letters from some alphabet. In such a case, we denote by face {1} the color {red} or the letter {B}, by face {2} the color {blue} or the letter {X}, and so on. In other words, the faces of the dice do not have to be a number between 1 and 6. However, for convenience, we shall always assume that the faces contain a number of dots between 1 and 6.

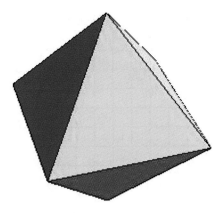

Fig. 3.1 A dice with eight faces.

Here are a few questions:

(i) Suppose that I have constructed a dice with eight faces (Fig. 3.1), each with different colors — red, blue, green, etc. What is the sample space of this dice?

(ii) What if a regular six-faced dice is colored with three colors? Two opposite faces are red, two opposite faces are blue and two opposite faces are green. What is the sample space of this dice?

(iii) You are shown an urn containing two identical red marbles, three identical blue coins and three identical yellow dice. You put your hand inside the urn and draw one and only one object from the urn. What is the sample space in this game?

(iv) A square board of size 1 m^2 is divided into small squares, each with an edge of 10 cm (Fig. 3.2). I throw a dart and tell you that it hit the board in one of these squares (we ignore the possibility that the dart hit a line bordering two squares or did not hit the board at all). What is the sample space in this game?

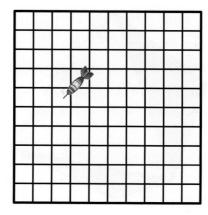

Fig. 3.2   A board divided into 100 (equal-area) regions.

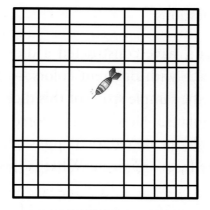

Fig. 3.3   A board divided into 100 unequal regions.

(v) The same board of size 1 m² is now divided into 100 regions of different sizes and forms (Fig. 3.3). Again, we throw a dart and we assume that it hits one and only one of these regions (ignoring a hit at one of the bordering lines). What is the sample space in this game?

Answer these questions before consulting Note 1.

Now that we have looked at a few examples of a sample space, we can define the sample space as a set of all possible outcomes of an experiment or a game. We denote by Ω the *set* of all possible outcomes and list the outcomes within the curly brackets { }. We assume that when we perform the experiment, one and only one of these outcomes will occur. We usually refer to these outcomes as *elementary events*. The word *elementary* is used here in the sense that these are the individual outcomes — one and only one of these occur. In some textbooks, the term "elementary event" is used for outcomes that have equal probability. We shall not use this meaning in this book. Here, *elementary events* are to be contrasted with *compound events*, which are discussed in the next subsection. It should be noted, however, that in many cases we are free to choose the events that we consider as elementary. We demonstrated this with the examples above.

Suppose we are told that we have a square board of size 1 m². We are told that a dart was thrown at the board and that it hit at some point in the region within the board (Fig. 3.4). Clearly, in this case there is an *infinite* number of points at which the dart could hit. The elementary events are all the *points* within this square. In this book, we shall never discuss an infinite sample space. This is done within the mathematical theory of

Fig. 3.4   A dart hitting a point on a square board.

probability. However, in practice, we are never interested in the exact mathematical point at which the dart hit. In practice, we always divide the board into some small regions, as in examples (iv) and (v). Once we have made this division and we are interested only in knowing in which of the small regions the dart hit, then we have a *finite* number of elementary events. Note, however, that there are many ways in which we can choose to make the division of all points into elementary events. Therefore, we must clearly describe the sample space we have chosen for each case.

For instance, consider a modification of example (iii) — an urn with 14 marbles. Suppose that all of the marbles are distinguishable, say they are numbered or they have some other labels that distinguish one marble from the other. In addition to these labels, they are also colored, say six are red, five are blue and three are yellow.

We do the same experiment as described above. What is the sample space? Clearly, there are at least two possible choices of the sample space depending on which outcome we consider to be an elementary event: "Draw a *specific marble*" or "draw a *specific color*." In the first choice we have 14 elementary events. In the second we have three elementary events. We shall return to this aspect of the choice of elementary events when we discuss the probabilities assigned to the elementary events.

3.3 The Set of All Possible Events: *L*

Once we have chosen the elementary events of an experiment, we can define *compound* events. A compound event is simply a *subset* of all the elementary events (a subset of a set is simply a set that includes a part or all of the elements of the set).

Consider again the case of a fair dice. The set of all possible elementary events is $\Omega = \{1, 2, 3, 4, 5, 6\}$. A compound event is any *subset* of Ω. For example:

$$\{\text{An even outcome}\} = \{2, 4, 6\}$$
$$\{\text{Larger than four}\} = \{5, 6\}$$
$$\{\text{Larger than four } or \text{ smaller than two}\} = \{1, 5, 6\}$$
$$\{\text{Larger than four } and \text{ smaller than six}\} = \{5\}$$
$$\{\text{Larger than five}\} = \{6\}$$
$$\{\text{Larger or equal to one}\} = \{1, 2, 3, 4, 5, 6\} = \Omega$$
$$\{\text{Smaller than one}\} = \{\ \} = \emptyset.$$

Here, $\emptyset$ denotes the empty set.

Clearly, the number of all possible compound events is much larger than the number of elementary events. Note also that an elementary event, say $\{6\}$, is included in the list of all compound events. For this reason, some textbooks refer to a compound event simply as an event.

In the last two examples, we have also added to the list of events the event Ω, which we shall refer to as the *certain event*, and the event $\emptyset$, which we refer to as the *empty event*. We always assume that an experiment is performed and one of the elementary events listed in Ω must have occurred. The last example is the empty event. There is no outcome that is smaller than one. We call this event the empty event or the impossible event, and we denote it by the symbol $\emptyset$. This is tantamount to saying that the event Ω did not occur. The impossible event and the certain event are sometimes referred to as *improper* events, and all other events are referred to as proper events, i.e. they consist of a *proper subset*, such that the number of its elements is greater than zero and smaller than the total number of elementary events.

Can you count how many events are in *L*? This is your first *mathematical challenge*. Let us work on this problem starting from the simplest sample space.

In all the following examples, we are given an urn with n labeled marbles. For simplicity, we use the integers $1, 2, 3, \ldots, n$ for the labels.

The case of one marble: $\Omega_1 = \{1\}$

In this case, we have one elementary event $\{1\}$. We have two kinds of events: the impossible event $\{\ \}$ and the certain event $\Omega_1 = \{1\}$. Altogether, we have *two* events in *L*.

The case of two marbles: $\Omega_2 = \{1, 2\}$

Here we have two proper subsets, either $\{1\}$ or $\{2\}$. Hence, we have *two* proper events. We also add the impossible event and the certain event. Altogether, we have *four* events in *L*. These are $\{\ \}, \{1\}, \{2\}$ and $\{1, 2\}$.

The case of three marbles: $\Omega_3 = \{1, 2, 3\}$

In this case, we have *one* impossible event and *three* events consisting of one elementary event each: $\{1\}, \{2\}$ and $\{3\}$. There are also another *three* events consisting of two elementary events each: $\{1, 2\}, \{1, 3\}$ and $\{2, 3\}$. Finally, there is *one* certain event: $\{1, 2, 3\}$. Altogether, we have *eight* events in *L*.

The case of four marbles: $\Omega_4 = \{1, 2, 3, 4\}$

In this case, we have *one* impossible event and *four* events consisting of one elementary event each. These are: $\{1\}, \{2\}, \{3\}$ and $\{4\}$. We have *six* events consisting of two elementary events each. These are: $\{1, 2\}, \{1, 3\}, \{1, 4\}, \{2, 3\}, \{2, 4\}$ and $\{3, 4\}$.

We have *four* events consisting of three elementary events each. These are: $\{1, 2, 3\}$, $\{1, 2, 4\}$, $\{1, 3, 4\}$ and $\{2, 3, 4\}$. Finally, we have *one* certain event: $\{1, 2, 3, 4\}$. Altogether, we have 16 events in *L*.

The case of five marbles: $\Omega_5 = \{1, 2, 3, 4, 5\}$

This requires a little more thinking and a little more writing, but no new ideas are involved. I urge you to count all of the possible events for this case. Write all the possible events consisting of zero, one, two, three, four and five elementary events.[2]

The case of six marbles: $\Omega_6 = \{1, 2, 3, 4, 5, 6\}$

You will recognize that this is equivalent to the case of a six-faced dice. Again, analyzing all possible cases require a little more work and more writing, but no new principles are involved. I suggest that you defer working on this case. Meanwhile, let us discuss the more general case. This will give us a general and a powerful tool to analyze the case Ω_6, as well as any other case (e.g. Ω_7, Ω_8 and so on).

The general case of an urn with *n* labeled marbles: $\Omega_n = \{1, 2, 3, \ldots, n\}$

This section deals with combinatorics — the art of calculating how many ways we can arrange *n* objects in *k* boxes. Although this is not essential for understanding the rest of the book, I urge you to read through it.

Again, there is only one impossible event. It is also easy to see that there are exactly *n* events, each of which consists of *one* elementary event. These are: $\{1\}$, $\{2\}$, $\{3\}$, $\ldots$, $\{n\}$.

Calculating the number of events consisting of two elementary events requires more work. Let us pause for a moment to construct two very useful tools for counting.

The notation $n!$ is defined by: $n! = 1 \times 2 \times 3 \times \cdots \times n$, i.e. multiply all integers from 1 to n. This number is referred to as *n-factorial*. It is also the number of *permutations* of n different objects. This is the number of ways we can order the n distinct objects. To count this number, suppose we have n boxes and we want to distribute the n objects into the n boxes. In the first box we can put any one of the n objects. In the second box, we can put any one of the $(n - 1)$ remaining objects. For each of these arrangements, we can choose one of the $(n - 2)$ remaining objects to put in the third box, and so on until we reach the last, or the nth box, for which we have only one object left.

Let us do the calculation for $n = 3$. Say three marbles having different colors: Red, blue and green. We can put one of the three colors in the first box. These are shown in Fig. 3.5a.

For each of these arrangements, we can put one of the two remaining marbles in the second box, so we get the six "configurations" shown in Fig. 3.5b.

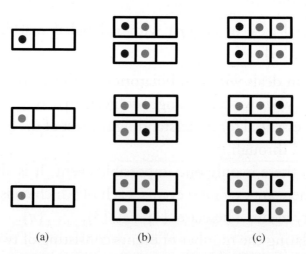

(a)   (b)   (c)

Fig. 3.5 (a) Three different configurations for the first box. (b) Six different configurations for the first and second boxes. (c) Six different configurations for the three boxes.

For the next step we have only one marble left, so we have only one choice: To put it in the third box (Fig. 3.5c). So altogether, we have $3! = 3 \times 2 \times 1 = 6$ possible permutations of three marbles in three boxes. These are shown in Fig. 3.5c. In general, we have $n!$ permutations to put n distinct objects into n ordered boxes.

How many permutations are there to put n marbles into k ordered boxes? This is called k-permutations of n things. As before, we can put one of the n marbles in the first box. For each of these arrangements, we can choose one of $(n - 1)$ marbles to put in the second box, $(n - 2)$ in the third box, and so on until we get to the last box, the kth box. At this stage, we have already placed $(k - 1)$ marbles, so we have in our hand $n - (k - 1) = n - k + 1$ marbles, and we can put one of these in the last box. Altogether, the number of k-permutations of n objects is:

$$P(n, k) = n(n - 1)(n - 2) \cdots (n - k + 1).$$

We can also rewrite $P(n, k)$ as a ratio of two factorials:

$$P(n, k) = n(n - 1)(n - 2) \cdots (n - k + 1) \times \frac{(n - k)!}{(n - k)!}$$

$$= \frac{n!}{(n - k)!}.$$

All we did was to multiply $P(n, k)$ by $(n - k)!$, and then divide by $(n - k)!$ to obtain the final ratio of the two factorials.

Check this formula for some small n and k of your choice.

Note that $1! = 1$, but $0!$ has not been defined. It is convenient to define this as $0! = 1$.

Note that for $n = k$, this is simply $P(n, n) = n!$

Exercise: Calculate $P(6, 2)$. First use the direct calculation of placing six objects into two boxes, then use the general formula as above with $n = 6$ and $k = 2$.[3]

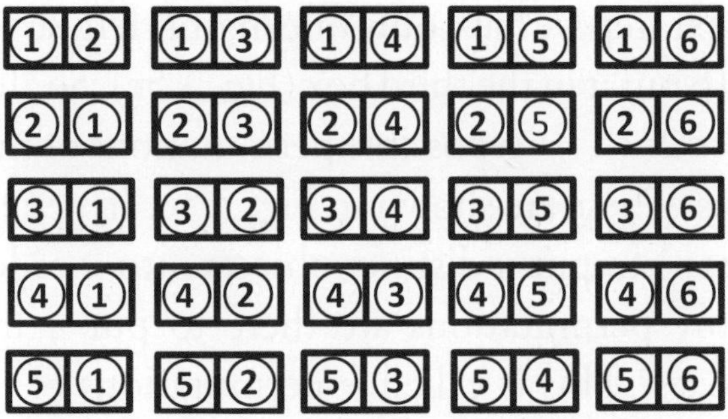

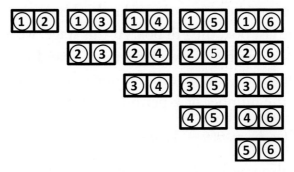

Fig. 3.6 All possible configurations for six labeled marbles in two boxes.

Fig. 3.7 All possible configurations for six labeled marbles in two boxes, disregarding the order of the boxes.

In the next quantity, we need to define the so-called *k*-permutations of *n* objects. This is the same as $P(n, k)$, but we disregard the *order* of the objects. In the example shown in Fig. 3.6, we see that if we disregard the *order* of the boxes, the choice of marble "1" in the first box and marble "2" in the second box is the same as the choice of marble "2" in the first box and marble "1" in the second box, and so on. Altogether, the number of *k*-combination is reduced from 30 to 15 (see Fig. 3.7).

In general, we define the *k*-permutations of *n* objects by:

$$C(n, k) = \frac{P(n, k)}{k!} = \frac{n!}{(n - k)!k!}.$$

This is also known as the *binomial coefficient* and is denoted by the symbol $\binom{n}{k}$.

Exercise: Suppose that in a class of 30 students we need to select a committee of two. One is to be the committee's head, and the other one is to be the deputy. Clearly, this problem is the same as counting $P(30, 2)$. We can replace the 30 students by 30 marbles, and the committee by two boxes. *The order is important* in terms of the first and the second choices (i.e. the head and the deputy). The number of choices is clearly 30×29. (There are 30 possibilities of choosing the head, and there are 29 possibilities left for choosing the deputy).

On the other hand, we might be interested in the number of ways of choosing a committee of two students without assigning a *specific order*, a specific function or a specific title to each selected student. Clearly, in this case we have $30 \times \frac{29}{2}$ possibilities. The division by 2 is a result of our disregarding the "order," i.e. the selection of Tom as head and Alice as a deputy or Alice as head and Tom as deputy are now the *same* committee. We simply need a committee of *two students*; this is the same as going from Fig. 3.6 to Fig. 3.7.

Now that we have the proper notations, let us go back to the general case: We are given an urn with n marbles. The marbles are labeled by numbers, letters or colors. How many different selections of k marbles ($k \leq n$) are there if we are not interested in the *order* of the selected marbles? This number is simply:

$$C(n, k) = \binom{n}{k} = \frac{n!}{(n-k)!k!}.$$

You do not have to memorize this formula. You should be able to *derive* it and re-derive it whenever it is needed. If you feel uncomfortable with this formula, go back to the examples of $n = 2, 3, 4, 5$ that we did above. Check each case with and without the formula and convince yourself that this formula is correct.

After you have done the case $n = 5$, let us calculate how many events are in Ω_6 for the $n = 6$ case.

There is one impossible event: $\emptyset$.

There are six events that consist of one elementary event: $\{1\}, \{2\}, \ldots$.

There are 15 events that consist of two elementary events: $\{1, 2\}, \{1, 3\}, \ldots$.

There are 20 events that consist of three elementary events: $\{1, 2, 3\}, \{1, 2, 4\}, \ldots$.

There are 15 events that consist of four elementary events: $\{1, 2, 3, 4\}, \{1, 2, 3, 5\}, \ldots$.

There are six events that consist of five elementary events: $\{1, 2, 3, 4, 5\}, \{1, 2, 3, 4, 6\}, \ldots$.

There is one event consisting of six elementary events: $\Omega = \{1, 2, 3, 4, 5, 6\}$.

We write these numbers as

$$1, 6, 15, 20, 15, 6, 1.$$

The total number of events in L is 64.

You will notice that the total number of events in each case is 2^n (e.g. for $n = 6$, we get $2^6 = 64$ events). Is this always true? To check this, and also prove that this is true, consider the general case of Ω_n, i.e. when the sample space is: $\Omega = \{1, 2, 3, \ldots, n\}$.

We want to count how many *subsets* of Ω we can form. A subset of two elementary events, say $\{2, 4\}$, is a subset of three elementary events, say $\{1, 5, 7\}$, and so on. Each of these examples may be written as a list of YES and NO responses, e.g.

$\{2, 4\}$ is the same as "1" is NO, "2" is YES, "3" is NO, "4" is YES, and so on. (YES means "present" in the subset.)

Each selection of a subset of this set may be written as a *list* of the form:

$$\{YES, NO, YES, YES, \ldots, NO\}.$$

The list of the "YES" and "NO" outcomes means that the object numbered 1 is present (YES), the object numbered 2 is not present (NO), objects 3 and 4 are present (YES, YES), and so on. Clearly, the total number of possible selections is the same as the total number of subsets of the set n objects. How many subsets of Ω can we form?

Each subset is a list of n entries of YES or NO. There are two possible entries (YES or NO) in the first position, two possible entries (YES or NO) in the second position, and so on. Therefore, there are $2 \times 2 \times 2 \times 2 \cdots \times 2 = 2^n$ possible lists with YES and NO entries. Specifically, the impossible event is represented by the list:

$$\{NO, NO, \ldots, NO\}$$

i.e. none of the objects is present (or no object was selected). On the other hand, the *certain* event is represented by the list:

$$\{YES, YES, \ldots, YES\}$$

i.e. all the objects are present, or were selected.

Now that we have the total number of subsets of the set of n objects, we can ask how many selections are there of sets of k objects. These numbers are:

$$\text{for } k = 0, \quad \binom{n}{0} = 1,$$

$$\text{for } k = 1, \quad \binom{n}{1} = n,$$

$$\text{for } k = 2, \quad \binom{n}{2} = \frac{n!}{2},$$

and so on,

$$\text{for } k = n, \quad \binom{n}{n} = 1.$$

Thus, for each n we have $n + 1$ numbers (note that we added the zero case):

$$\binom{n}{0}, \binom{n}{1}, \binom{n}{2}, \ldots, \binom{n}{n}.$$

These numbers can be arranged in a triangle.

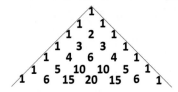

We have added the case $n = 0$, which is of no interest (i.e. it has no elementary outcomes) to obtain this triangle, which is known as the *Pascal triangle*. This triangle has many nice and interesting properties. It is very easy to construct. At each level, simply sum the two numbers right above it. For instance, the number 2 on the second line is obtained from $1 + 1$, the number 3 on the third line is obtained as either $(1 + 2)$ or $(2 + 1)$, etc.

We also found that the *total* number of selections is 2^n. Hence, we have the important equality:

$$2^n = \binom{n}{0} + \binom{n}{1} + \binom{n}{2} + \cdots + \binom{n}{n}.$$

Thus, we have discovered a particular case of the Newton binomial theorem.[4] You need not memorize this beautiful equation, but it is a very useful one. I urge you to try to calculate all of these numbers for a few n.

3.4 The Probability Function: *P*

So far, we have discussed two elements of the probability space (Ω, L, P). One is the set of all possible outcomes of an experiment, say 1, 2, 3, 4, 5 or 6 in throwing a dice. The second is all possible "events," e.g. $\{1, 6\}, \{1, 2, 3\}$, etc. This section introduces the third element: The *probability function*, denoted *P*. You might justifiably ask: We have already discussed the concept of probability in Sessions 1 and 2, why "introduce" this concept again here?

The answer to this question is not simple. In the first two sessions, we tried to "define" the concept of probability. In some textbooks, along with the classical and the frequency "definitions" of probability, you can also find a third "definition" of probability, which is referred to as the "axiomatic definition." As we have seen in Sessions 1 and 2, neither the classical nor the frequency "definitions" are *bona-fide* definitions of probability. The same applies to the so-called axiomatic definition.

In the axiomatic development of the theory of probability, one *introduces* the probability function as a *measure* of the "size" of the events without any discussion of its meaning or of the method of its calculation. The axiomatic approach does not *define* the concept of probability but assumes that such a function exists, and has certain properties. These properties are:

(a) The range of values of the probability function is: $0 \leq \Pr \leq 1$.
(b) The certain event is assigned the probability one: $\Pr(\Omega) = 1$.

(c) If A and B are disjoint events, then:

$$Pr(A \ or \ B) = Pr(A) + Pr(B).$$

Two events are said to be "disjoint" when they do not have any elementary events in common. Another definition of two disjoint events is that the occurrence of one event excludes the possibility of occurrence of a second event. Some examples are given below.

The motivation for assuming these properties is based on our expectation of how a measure of the "size" of the events should behave. First, we assume that a measure of the size of the event exists. Second, we choose the range of values that such a measure can attain. This range is usually chosen to be between zero and one, i.e. $0 \leq Pr \leq 1$. In daily discussions of probabilities, one usually chooses percentages as a reference, i.e. between 0% and 100%.

We assign the probability *zero* to an *impossible* event and probability *one* to a *certain* event. These assignments determine the *range* of values that the probability function can attain.

The next property we assume for the probability function is called the *additivity*. Clearly, if the probability function is supposed to measure the *size* of the event, then the "larger" the event, the larger should be its size, and hence the larger its probability.

But what do we mean by a "larger" or "smaller" event?

Consider the following sentences that you might hear people saying:

(i) Yesterday, there was a gala performance of the new opera. This event took almost 4 hours, far *longer* than any other concert I have ever watched.

(ii) The Second World War took the lives of millions — a far *greater* number than any other preceding war.

(iii) Yesterday, I got married and it was the *greatest* event of my life.

In all of these, and many other sentences, we quantify events according to some criteria. In (i) the event is "greater" in the sense that it took a *longer* time. In the second case (ii) it is "larger" in the sense that the event involved a larger number of casualties. In the third case (iii) the "size" of the event is a highly subjective expression of the extent of importance of that event to the person concerned.

In probability theory, the measure we assign to events is not the kind of measure that is expressed in the above examples. It is true that we could have said that the opera in example (i) was a rare event, and that the probability of its occurrence is, say, 0.1%. Similarly, we could have assigned probabilities to the occurrence of wars, weddings or any other events. In probability theory, we assign probabilities to events that are outcomes of well-defined experiments, and have well-defined numbers of elementary outcomes. Let us now "discover" what we mean by the additivity of the probability function.

Consider first a fair dice having six outcomes: $\{1, 2, 3, 4, 5, 6\}$. Suppose you can bet on one of the following events:

$$A: \{1, 2\}$$
$$B: \{1, 2, 3\}.$$

Event A means that you can bet that the outcome will be either "1" or "2." Similarly, event B means the outcome will be either "1," "2" or "3."

Without giving a numerical *value* for the probabilities of these events, which event do you believe has a higher chance of winning?

Next, consider these two events:

$$C: \{1, 2\}$$
$$D: \{4, 5, 6\}.$$

Which of these has a larger chance of winning?[5]

After making your guesses on these two examples, consider again the same two examples but for a situation in which the dice is not fair, i.e. the elementary events are not equally likely to occur. Which of the events A or B will you bet on? Which of the events C or D will you bet on?

Can you tell me why I have chosen these particular examples?

Before we discuss the different probabilities of the events, it should be clear that if the dice is fair, i.e. each of the elementary events has the same probability or likelihood of occurrence, then the event consisting of a larger number of elementary events must have a larger probability. Thus, in this case, event B is "larger" than event A, and therefore has a higher probability. Similarly, event D is "larger" than event C, and therefore must have a larger probability.

Note also that the "larger" event in cases A and B, or in C and D, means a larger *number* of *elementary events*. B contains more elementary events than A, and D contains more elementary events than C. However, there is an important difference between the two examples. In the first example, event B *includes* event A. In other words, A is a subset of B. In set theory, this is denoted $B \supset A$, or $A \subset B$. It is always true that an event that is a subset of another event has a smaller probability, i.e. from $B \supset A$, it follows that $Pr(B) > Pr(A)$. (This is true provided the probabilities of all elementary events are not zero.) Construct some simple examples and convince yourself that this is always true.

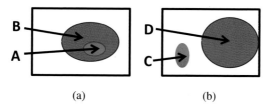

(a)   (b)

Fig. 3.8   In (a) **A** is smaller than **B**, and **A** is a subset of **B**. In (b) **C** is smaller than **D**, but **C** is not a subset of **D**.

In the second example, D is larger than C because it contains more elementary events. However, C is not a subset of D, i.e. C is not *included* in D.

Figure 3.8 demonstrates the difference between the two cases in terms of regions on a board. This kind of presentation is referred to as a Venn diagram. In Fig. 3.8a, A is a subset of B. It is also clear that the probability of hitting B with a dart that is thrown blindly at the board is larger than hitting A. In Fig. 3.8b, region D is larger than region C (in the sense that it contains more points on the board). Here, the probability of hitting D is larger than the probability of hitting C. However, in this case, C is *not* a subset of D.

We conclude that for the case of a *fair* dice (or any other experiment having equally probable elementary events), a larger event has a larger probability, with "larger" here being in the sense of having more elementary events. This is true whether or not the larger event *contains* the smaller event (or whether the smaller event is a subset of the larger event). This conclusion does not hold for elementary events that are not equally probable.

Now consider an unfair dice. Here, we do not know anything regarding the *probability distribution*, i.e. we do not know the values of $Pr(1), Pr(2), \ldots, Pr(6)$.

In this case, event A is smaller than event B, and event C is smaller than event D, with "smaller" here being in the sense that it contains fewer elementary events. In both examples, the "smaller" events (A and C) have two elementary events each, and the "larger" events (B and D) have three elementary events each. However, the probabilities we assign to the various events depend on the probabilities of the various elementary events. If the dice is fair, each elementary event is assigned a probability 1/6. Clearly, then, events A and C will have probability 2/6. Events B and D will have probability $3/6 = 1/2$. On the other hand, suppose that you are given the following probabilities:

$$Pr(1) = \frac{1}{4}, \quad Pr(2) = \frac{1}{4},$$

$$Pr(3) = Pr(4) = Pr(5) = Pr(6) = \frac{1}{8}.$$

In this case, we have the probabilities:

$$Pr(A) = \frac{1}{4} + \frac{1}{4} = \frac{1}{2}$$

$$Pr(B) = \frac{1}{4} + \frac{1}{4} + \frac{1}{8} = \frac{5}{8}$$

$$Pr(C) = \frac{1}{4} + \frac{1}{4} = \frac{1}{2}$$

$$Pr(D) = \frac{1}{8} + \frac{1}{8} + \frac{1}{8} = \frac{3}{8}.$$

Thus, we see that for this unfair dice, $Pr(A)$ is also *smaller* than $Pr(B)$. However, $Pr(C)$ is *larger* than $Pr(D)$. Check carefully the calculations I have carried out above and convince yourself that these results are plausible. Why does event C, which has fewer elementary events, have a larger probability compared with event D?

There is an important conclusion that we can draw from these examples. We must make a distinction between the *number* of elementary events contained in an event and the *size* of that event. If the elementary events are equally probable (e.g., a fair dice), then the larger the number of elementary events, the larger the size of the event, and the larger the probability. The last statement is not true when the probability distribution is *non-uniform*, i.e. when the probabilities of the elementary events are not equal. As we saw in the examples above, a larger *number* of elementary events does not guarantee a larger probability.

While calculating the probabilities of the events A, B, C and D, we intuitively used the *sum rule*, which means that the probability of any compound event is simply the sum of the probabilities of the elementary events included in that event.

We now generalize the sum rule for the general event.

First, we define two disjoint events (or mutually exclusive events) as two events that do not have any elementary events in common. For instance, C and D in Fig. 3.8 are disjoint events. The occurrence of one event excludes the possibility that the second event occurred. Similarly, events C and D for the dice defined above are disjoint events.

What is the probability of *either* event C *or* D occurring, i.e. either $\{1, 2\}$ or $\{4, 5, 6\}$?

This is the same as the probability of the compound event $(1, 2, 4, 5, 6)$, which is:

$$Pr(1, 2, 4, 5, 6) = Pr(1) + Pr(2) + Pr(4) + Pr(5) + Pr(6)$$
$$= Pr(1, 2) + Pr(4, 5, 6).$$

We can conclude that the *probability* of *two disjoint events* is the *sum* of the *probabilities* of the *two events*.

We have arrived at this statement simply by applying the sum rule to the elementary events. We listed all of the elementary events in the event {either C or D}, and simply summed over all of the probabilities of the elementary events. We use the symbol $\cup$ for the *union* of two events, i.e. $A \cup B$ means *either* A *or* B (or both) has occurred. For any two disjoint events, the rule we have discovered can be written as:

$$Pr(\mathbf{C} \text{ or } \mathbf{D}) = Pr(\mathbf{C} \cup \mathbf{D}) = Pr(\mathbf{C}) + Pr(\mathbf{D}).$$

We also use the symbol $\cap$ for the *intersection* of the two events, i.e. $A \cap B$ means *both* A *and* B occurred. In the example above, we have $A \cap B = \{1, 2\}$. This means that events A *and* B, which is $\{1, 2\}$, has occurred. On the other hand, $C \cap D = \emptyset$, which means that both C and D have occurred. Clearly, this is an impossible event.

The corresponding probabilities are:

$$Pr(A \cap B) = Pr\{1, 2\} = Pr(1) + Pr(2)$$
$$Pr(C \cap D) = Pr(\emptyset) = 0.$$

Now calculate the probability of the union of two non-disjoint events, say $E = (1, 2, 3)$ and $F = (3, 4)$.

What is the probability or the union $E \cup F$, $Pr(E \cup F)$?

To calculate the probability of $E \cup F$, we cannot use the sum rule, since E and F are not disjoint. However, we can use the sum rule applied to the *elementary events*, i.e.

$$Pr(E \cup F) = Pr\{1\} + Pr\{2\} + Pr\{3\} + Pr\{4\} = 4/6.$$

Note that in this "counting" of the elementary events, we counted the event $\{3\}$ only *once*, although it appears in both E and F.

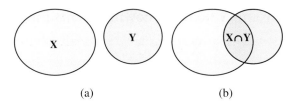

Fig. 3.9 (a) Non-overlapping events, (b) overlapping events.

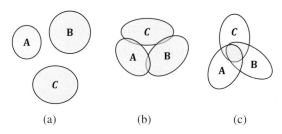

Fig. 3.10 (a) Three disjoint events. (b) Each pair of events overlap. (c) The three events overlap.

Thus, we can arrive at the sum rule for any two events X and Y:

$$Pr(X \cup Y) = Pr(X) + Pr(Y) - Pr(X \cap Y).$$

If X and Y are disjoint, then $Pr(X \cap Y) = 0$. Hence, the sum rule in this form reduces to the rule we had for disjoint events. The reason we must subtract $Pr(X \cap Y)$ is that when we count all the elementary events in X and in Y separately, we count the common event $(X \cap Y)$ twice. See also Fig. 3.9 for the corresponding Venn diagram of (a) *disjoint* events and (b) *overlapping* events. Convince yourself that the probability of hitting the area $(X \cup Y)$, i.e. hitting either X or Y or both, is the sum of the probabilities of hitting X and Y in case (a), but it is not the sum of hitting X and Y in case (b), where we must subtract the probability of the overlapping area, i.e. of $Pr(X \cap Y)$.

Exercise: Let X and Y be the following events for a fair dice:

$$X = \{1, 2, 3, 4\}$$
$$Y = \{2, 4, 6\}$$

Calculate $Pr(X)$, $Pr(Y)$ and $Pr(X \cup Y)$.[6]

Further exercises:

1. Three events A, B and C are disjoint, i.e. there are no elementary events that are common to the three events A, B and C. Does it follow that each *pair* A, B or B, C or A, C consist of two disjoint events? Give examples.
2. Given that A and B are disjoint and B and C are disjoint, does it follow that A, B and C are disjoint? (A, B and C are said to be disjoint if there are no elementary events that are common to all the three events A, B and C.)

Can you generalize the sum rules for three events?[7]

Before ending this session, let us take a look at one more experiment that was conducted on children in grades 5 to 11.[8]

The following question was posed to children: Suppose there are 10 students and we want to select two committees:

(i) A committee of two students (out of 10 and independent of the order).
(ii) A committee of eight students (out of 10 and independent of the order).

The questions is the following: in which case is the number of possibilities larger — in (i) or in (ii)?

The answers that the children gave were:

(a) (i) is smaller than (ii);
(b) (i) is equal to (ii);

(c) (i) is greater than (ii);
(d) Other;
(e) No answer.

The percentages of each given answer are shown in Table 3.1 below.

Table 3.1

Answer	Student Grade				College Students
	5	7	9	11	
(a)	20	5	10	0	22
(b)	0	5	5	15	6
(c)	10	20	65	85	72
(d)	15	30	15	0	0
(e)	55	40	5	0	0

The correct answer is (b). If you are not sure of this, calculate the number of possibilities in (i) and (ii) and you will find that it is $10 \times \frac{9}{2} = 45$. It is interesting to note that the answer (c) is a common misconception.

3.5 Summary of What we have Learned in this session

We *define* the probability space by the triplet (Ω, L, P), where Ω is the set of all possible outcomes of an experiment (or a game or a trial); L is the set of all possible events (or compound events); and P is the probability function, which is defined for each of the events in L. It assigns a positive number $0 \leq Pr \leq 1$ to each event, such that $Pr = 0$ is assigned to an impossible event and $Pr = 1$ is assigned to a certain event. This function has the sum property:

$$Pr(A \cup B) = Pr(A) + Pr(B) - Pr(A \cap B)$$

which, for disjoint events, reduces to:

$$Pr(A \cup B) = Pr(A) + Pr(B)$$

Note carefully that, in defining the probability space, we introduce the function Pr, but we did not discuss the meaning of the probability of an event, nor the method we used to measure or calculate the probability of an event.

Session 4

Independence and Dependence Between Events

Up to this point, we have discussed sets of events and a measure (probability) assigned to these events. We now arrive at one of the most important concepts in the theory of probability: *Conditional probability*. This concept is central to the theory of probability and does not feature in set theory. It is this concept that makes probability theory more useful, more interesting and more exciting than set theory. This is also the reason why this session is the longest one in this book.

As a quick warm-up, consider the following easy problems:

(a) You are shown two urns. In each of the urns, there are two red marbles and two blue marbles (Fig. 4.1). What is the probability of picking a red marble from the left urn?

What is the probability of picking a red marble from the right urn?

What is the probability of picking a red marble from the left urn *and* a red marble from the right urn?

(b) You are shown one urn containing two red marbles and two blue marbles.

What is the probability of picking a red marble?

87

Left Urn　　　　　　**Right Urn**

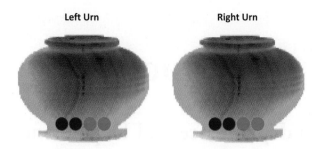

Fig. 4.1　Two urns with four marbles; two red and two blue.

Suppose you pick a red marble in the first trial. You return the marble to the urn, shake the urn and try once again to pick a marble.

What is the probability of picking a red marble on the second trial?

(c) As in (b) you start with the same urn, having two red marbles and two blue marbles.

You pick one red marble in the first trial. The probability of this event is 1/2.

You *do not* return the marble to the urn. What is the probability of picking a red marble in the second trial given that you have already drawn a red marble in the first trial, but the marble was not returned to the urn?[1]

If you answered the questions in (a), (b) and (c) correctly, you already have a sense of what independence or dependence between events mean. In fact, not only do you have a sense of these concepts, but you have also correctly calculated a *conditional probability*.

In general, two events are considered to be *independent* when the occurrence of one event has no effect on the probability of occurrence of the other event. For instance, throwing two dice at two different places are independent experiments, and their outcomes are independent events. On the other hand, if the

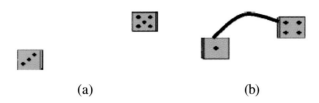

Fig. 4.2 (a) Independent and (b) dependent dice.

two dice are tied by some inflexible wire (Fig. 4.2), then the outcome on one dice affects the probability of the outcome on the second dice.

Similarly, if you draw a marble from two urns, then the outcomes are independent. The same is true if you draw a marble from an urn, then return the marble to the urn, and draw again. The two consecutive experiments are independent again.

On the other hand, if you draw a marble, see the result and do not return it back into the urn, then clearly the probabilities of the outcomes on the second draw would depend on the outcome of the first draw. Thus, in example (c) above, *given* that you drew a red marble, the probability of drawing a red marble again is 1/3, and the probability of drawing a blue marble in the second instance is 2/3.

The concept of dependence is also clear for events that we encounter in daily life. Suppose that it is raining in Jerusalem today. This fact has (almost) no effect on the probability of it raining during the next day in New York. However, the fact that it is raining in Jerusalem today might affect the probability that it will rain in Jerusalem the next day. This is of course a very intuitive and qualitative assessment. We shall be interested in a more precise definition and methods of calculating probabilities of dependent events.

Suppose I throw a dice and I tell you that the result is "even." What is the probability that the result is {4}? Note here that you have been given some *information* on the result. It is not the answer to the question we asked, but it is a "hint" that might help us find the answer. We write the required probability as:

$$Pr(\{4\} \mid given \ that \ the \ results \ is \ ``even")$$

or, for short:

$$Pr(\{4\} \mid ``even").$$

In this particular example, you should be able to calculate this *conditional* probability, assuming that the dice is fair and that you know that the result is "even." This means that the outcome could be either {2}, {4} or {6}. Each of these has the same probability of *one* out of *three*. Hence, the answer to the question is 1/3.

Note at this point that, when I ask you what the probability of {4} is, giving you only the information that the dice is fair, the required probability can also be written as a conditional probability:

$$Pr(\{4\} \mid given \ that \ the \ one \ of \ the \ results \ between \ 1 \ and \ 6 \ occurred).$$

We usually refer to the last probability as simply the *probability* of {4}, omitting in the notation that one of the results between 1 and 6 occurred.

Now that you have an intuitive feeling for what independence and dependence between events mean, let us try to find the general rule of calculating the probability of occurrence of both events A *and* B. This is often called the *intersection* of the two events, or the *product* of the two events. The reason for these terms will soon become clear. For the moment, we shall simple say {A *and* B}.

Suppose we throw two identical dice that are both fair, and they are far away from each other, so that you expect that the result on one dice will not affect the probability of the result on the other.

Define the two events:

$$A = \{First\ dice\ shows\ 4\}$$
$$B = \{Second\ dice\ shows\ 6\}.$$

We are interested in the probability of the event:

$$\{A\ and\ B\} = \{First\ showed\ 4\ and\ second\ showed\ 6\}.$$

A plausible reasoning to obtain the probability of these events is as follows: the specific result $\{A\ and\ B\}$ is an *ordered* pair of numbers (the order is important — the number on the left is the outcome of the *first* dice, and the number on the right is the outcome on the *second* dice). We write the required event as $\{(4, 6)\}$.

Note carefully that the notation $\{(4, 6)\}$ is the event: result "4" on the first dice *and* the result "6" on the second dice. This notation is different from the one we used in the previous sections, in which we denoted the event: "either "4" *or* "6" occurred on throwing a *single dice*" by $\{4, 6\}$.

Altogether, there are 36 different pairs of outcomes, which are shown in Table 4.1 below:

Table 4.1   All 36 Possible Results for Two Dice

(1, 1)	(1, 2)	(1, 3)	(1, 4)	(1, 5)	(1, 6)
(2, 1)	(2, 2)	(2, 3)	(2, 4)	(2, 5)	(2, 6)
(3, 1)	(3, 2)	(3, 3)	(3, 4)	(3, 5)	(3, 6)
(4, 1)	(4, 2)	(4, 3)	(4, 4)	(4, 5)	(4, 6)
(5, 1)	(5, 2)	(5, 3)	(5, 4)	(5, 5)	(5, 6)
(6, 1)	(6, 2)	(6, 3)	(6, 4)	(6, 5)	(6, 6)

We assume that the two dice are fair and that all of the outcomes on the two dice are independent. Therefore, we argue that each event in Table 4.1 has the same probability, i.e. one out of 36. Therefore, the probability of the required event is $Pr\{(4, 6)\} = \frac{1}{36}$. Here, we use the double brackets to emphasize that the event is "4" on the first and "6" on the second dice. In the following, we shall not use these double brackets when the event we are discussing is clear; we shall simply use the notation $(4, 6)$ instead of $\{(4, 6)\}$.

We know that the probability of obtaining a "4" on the first dice is 1/6. We also know that the probability of obtaining a "6" on the second dice is 1/6. We saw that the probability of the required event is:

$$Pr(4, 6) = \frac{1}{36} = \frac{1}{6} \times \frac{1}{6} = Pr(4) \times Pr(6).$$

The last formula is suggestive. Before we accept it as a general formula for any two independent events, let us try one more example of two independent events. We throw a fair dice and a fair coin simultaneously. We assume that the sample space of the dice has six outcomes:

$$\Omega_D = \{1, 2, 3, 4, 5, 6\}$$

And that the sample space of the coin has two outcomes:

$$\Omega_C = \{H, T\}$$

What is the probability of the event "outcome 4" on the dice *and* the "outcome H" on the coin? The sample space of the joint experiment is written below:

$$(1, H), (2, H), (3, H), (4, H), (5, H), (6, H)$$
$$(1, T), (2, T), (3, T), (4, T), (5, T), (6, T).$$

Altogether, there are 12 events, and if both the dice and the coin are fair and they were thrown at different locations so that one throw does not "know" about the other, it is reasonable to assume that all of these 12 outcomes are equally probable, and each has probability 1/12. Thus:

$$Pr(4, H) = \frac{1}{12} = \frac{1}{6} \times \frac{1}{2} = Pr(4) \times Pr(H).$$

Again, we see that the probability of the event "4" *and* "T" is the *product* of probabilities of the separate events "4" and "T".

Having these two examples in mind, we can generalize to any two independent events. First, we denote by $A \cap B$ the event "A *and* B." Sometimes, because of the product rule above, one uses the notation $A \cdot B$ for the event $A \cap B$. This notation might be confusing because we did not define a "product" of the two *events*. The product of the two probabilities in the above formula is a product of two numbers, whereas the notation $A \cdot B$ is not a product of two numbers. Therefore, we shall use either the notation A *and* B, or $A \cap B$.

The general rule for any two independent events is:

$$Pr(A \cap B) = Pr(A) \times Pr(B).$$

This rule is valid provided we know that the two events are independent. In most cases, we do not *know* whether or not the events are independent; therefore, the above formula is taken as a *definition* of independence. Two events are said to be independent if and only if the probability of the event $(A \cap B)$, i.e. (A *and* B), is the product of the probability of A and the probability of B.

Note carefully that in the two examples we worked through above, the events were elementary outcomes; a throw of a dice, or a toss of a coin. The rule above applies to any two independent events A and B, where A and B could be also compound events.

Also note that the formula above not only *defines* independence between two events, it also gives us the method on how to calculate the probability of the event $(A \cap B)$, i.e. that *both* A *and* B occurred.

It is important to distinguish between the notions of *disjoint* events and *independent* events. We say that two events are disjoint when they do not have any elementary events in common. Examples are: A = {even} and B = {odd}, i.e. an "even" result on one dice and an "odd" result on the *same* dice. Clearly, if A occurs, it means that B did not occur and vice versa. In general, two events are said to be disjoint when the occurrence of one of them excludes the possibility of the occurrence of the other.

Equivalently, we write for two disjoint events: $A \cap B = \emptyset$, where $\emptyset$ is the empty event, i.e. the impossible event. Thus, two events A and B are disjoint if the intersection between the two is empty. Therefore, the probability of the occurrence of $(A \cap B)$, i.e. (A *and* B), is zero.

Note, however, that the event "even" on one dice and the event "odd" on a second dice are not necessarily disjoint. Both events can occur, and if they occur on two fair dice that are far apart, the probability would simply be $\frac{1}{2} \times \frac{1}{2} = \frac{1}{4}$.

The notion of *disjoint* is defined in terms of the *events themselves*. For instance, the following two events for the same dice are *not* disjoint:

$$A = \{1, 2, 3\}$$
$$B = \{3, 4, 5, 6\}.$$

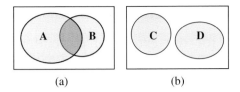

(a)   (b)

Fig. 4.3   (a) Overlapping events; (b) disjoint events.

On the other hand, the two events

$$C = \{1, 2\}, \quad D = \{5, 6\}$$

on the same dice are disjoint.

Similarly, in Fig. 4.3, the event "hitting region A" and "hitting region B" are not disjoint events. On the other hand, "hitting region C" and "hitting region D" are disjoint events.

In all of the examples above about *disjoint* events, we did not mention the probabilities of the events. This notion is a property of the events themselves and not of their probabilities. On the other hand, the notion of *dependence* and *independence* are *defined* in terms of the probabilities of the involved events.

When two events A and B are *not* disjoint, it means that they have some elementary events in common. For instance, the two events (on the same dice) $\{1, 2\}$ and $\{4, 5\}$ are disjoint. On the other hand, the following events (on the same dice) $\{1, 2, 3\}$ and $\{3, 4, 5\}$ are not disjoint. They have one elementary event in common: $\{3\}$.

The following events (on the same dice) have more elementary events in common: $\{1, 2, 3, 4\}$ and $\{3, 4, 5, 6\}$. Similarly, the events A and B in Fig. 4.3a are not disjoint. They have common points (shown in dark grey), but events C and D in Fig. 4.3b are disjoint.

We next turn to a measure of the *extent* of the *dependence* between two events.

4.1 Correlation Between Events

To get a feeling for the extent of dependence between two events, let us go back to the two different dice that are thrown simultaneously, but with one modification.

Suppose that each of the dice includes a small magnet at its center (Fig. 4.4). For the sake of correctness, assume that the north pole of the magnet points to the face with one dot, as shown in Fig. 4.4. We also assume that the magnet is very small and is made of the same material as the dice so that the dice is still a fair dice, i.e. the probability of any elementary outcome is 1/6. It should be said that we assume that the dice is thrown in a location in which there is no magnetic field, and it does not fall on some magnetic material. If the magnetic field of the earth is strong enough, it will obviously affect the orientation of the dice while it swirls in the air, so we exclude any external magnetic field.

On the other hand, for the purpose of studying dependence between events, we do assume that the magnet on one dice can affects the orientation of the second dice. This magnetic interaction depends on the distance between the two dice. The more distant they are, the weaker the interaction and the lesser the effect of one dice on the other. The maximal *attractive* interaction occurs when the south–north arrows of the two magnets are in opposite directions. The maximal repulsion occurs when these arrows are in the same direction (Fig. 4.5).

Fig. 4.4   A dice with a small magnet at its center.

Maximal attraction   **Maximal repulsion**

Fig. 4.5 Two extreme cases of orientations of the two magnets.

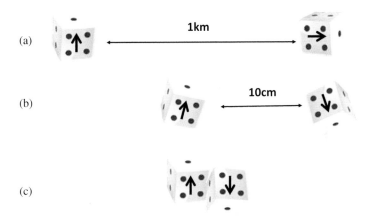

Fig. 4.6 Experiments with two dice at different distances.

Now we do several experiments. We throw the two dice simultaneously, but in each experiment we throw them at different distances.

(a) The two dice are thrown at a distance of 1 km apart (Fig. 4.6a). We expect that the outcomes on the two dice will be independent. If we want to be sure, we may "verify" this assumption by simply repeating the experiment many times. We find that each pair of results, say "1" on the first dice and "1" on the second dice, occurs with frequency of 1/36, which means that the probability of the event "A *and* B" is:

$$Pr\,(A \cap B) = \frac{1}{36} = \frac{1}{6} \times \frac{1}{6} = Pr(A) \times Pr(B).$$

As before, we can conclude that, at this distance, the two events A and B are independent. At this distance, there is no reason to believe that the magnet in one dice interacts with the magnet in the second dice, or if there is some interaction, it is negligible.

(b) We next repeat the same experiment but we throw the two dice simultaneously at a distance of about 10 cm from each other (Fig. 4.6b). At this distance, we repeat the experiment many times and find that the fraction of occurrence of "1" on the first dice and "1" on the second dice is not $\frac{1}{36}$, but slightly lower, say $\frac{1}{38}$.

We write this as:

$$Pr(1, 1) = \frac{1}{38} \neq Pr(1) \times Pr(2) = \frac{1}{6} \times \frac{1}{6} = \frac{1}{36}.$$

We say that there is a small *negative correlation* between these two events. On the other hand, we might find that the event "1" on the first dice and "6" on the second dice is slightly larger than $\frac{1}{36}$, say $\frac{1}{34}$. We write this as:

$$Pr(1, 6) = \frac{1}{34} \neq Pr(1) \times Pr(6) = \frac{1}{36}.$$

In this case, we say that there is a small *positive correlation* between the two events. The terms *positive* and *negative correlations* are due to the fact that in the first case:

$$Pr(1, 1) - Pr(1) \times Pr(1) = \frac{1}{38} - \frac{1}{36} < 0$$

whereas in the second case:

$$Pr(1, 6) - Pr(1) \times Pr(6) = \frac{1}{34} - \frac{1}{36} > 0.$$

We shall further discuss correlations between events below.

(c) Next, we repeat the same experiment, but we throw the two dice together so that they are at a very small distance from each other. While in the air, they might even stick to each other in such a way as shown in Fig. 4.6c. In this case, if we repeat the experiment many times (the distances between the two dice is always very small — in some cases they fall together, while in some other instances they fall at a short distance from each other), we may find that the fraction of results showing $(1, 1)$ is about $Pr(1, 1) = \frac{1}{150}$, i.e. this event is a very rare event. The correlation in this case is the difference between the probability of the event $(1, 1)$ and the product of the probability had the two events been independent, i.e.

$$Pr(1, 1) - Pr(1) \times Pr(6) = \frac{1}{150} - \frac{1}{36} < 0.$$

We see that, in this case, there is a large negative correlation between these two events. We can easily rationalize the reason for such a result. Because of the short distance between the two dice, the magnets will interact strongly, thereby tending to align themselves in such a way that the north pole of one magnet will be closest to the south pole of the second magnet. This is equivalent to aligning the two dice in such a way that the face "1" of one dice will be in the same direction as the face "6" on the second dice.

Indeed, if we collect the data on the occurrence of the event $(1, 6)$, we should find that the fraction of the occurrence of this event is much larger than the expected fraction had the two dice been independent. At short distances, the probability of the event $(1, 6)$ might be:

$$Pr(1, 6) = \frac{1}{5}$$

and the corresponding correlation:

$$Pr(1,6) - Pr(1) \times Pr(6) = \frac{1}{5} - \frac{1}{36} > 0$$

i.e. a very large positive correlation. The rationale for this result is the same as the one for the result $(1, 1)$ given above.

Exercise: Suppose you carry out the same experiment as in case (c) above. The two dice always get stuck to each other while they are in the air, and they remain stuck to each other as they fall, either one on top of the other or side by side, as in Fig. 4.7.

Fig. 4.7   Two dice with very strong magnets.

Can you estimate the *sign* of the correlation between the following pairs of events?[2]

$$(1, 4), (1, 6), (6, 2), (4, 4).$$

Exercise: You throw two coins many times and record the fraction of times each outcome occurs. When the two coins are thrown far away from each other, you find the following fractions of occurrence of the four events:

Outcomes: (H, H) (T, T) (H, T) (T, H).

Probabilities: $\frac{25}{100}$ $\frac{27}{100}$ $\frac{23}{100}$ $\frac{25}{100}$.

Fig. 4.8 Two coins with embedded magnets, one pointing towards T and the other pointing towards H.

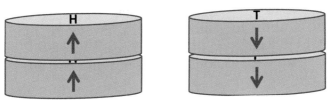

Fig. 4.9 The two most probable outcomes of the two coins with the magnets (see Note 3).

On the other hand, when they are thrown very close to each other, you find the results:

Outcomes: (H, H) (T, T) (H, T) (T, H).

Probabilities: $\frac{49}{100}$ $\frac{46}{100}$ $\frac{1}{100}$ $\frac{4}{100}$.

What would you conclude from these two sets of experiments (see also Fig. 4.9)?[3]

Exercise: Two marbles, one red and one blue, are placed in a rectangular box. The box has two compartments of equal size and they are large compared with the size of the marbles. The two compartments are not separated by a physical partition. The box is shaken vigorously, and while shaking, a camera takes snapshots of the locations of the marbles inside the box. Out of 100 pictures, you find the following fractions of events (Fig. 4.10):

(a) Two marbles on the left side: $\frac{40}{100}$.

(b) Two marbles on the right side: $\frac{42}{100}$.

(c) One red marble on the right and one blue marble on the left: $\frac{10}{100}$.

(d) One blue marble on the right and one blue marble on the left: $\frac{8}{100}$.

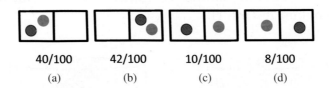

| 40/100 | 42/100 | 10/100 | 8/100 |
| (a) | (b) | (c) | (d) |

Fig. 4.10 Two marbles in two compartments.

Which results would you expect if the locations of the marbles were not correlated, i.e. if each marble had the same probability of being in the right or the left half of the box, and that they were independent?

What can you conclude about the dependence between the locations of the two marbles?

Can you estimate the correlation?[4]

Exercise: Suppose you perform the same experiment as in the previous exercise and you get the following fractions.

$$Pr(a) = \frac{5}{100}, \quad Pr(b) = \frac{4}{100}, \quad Pr(c) = \frac{45}{100}, \quad Pr(d) = \frac{46}{100}.$$

What can you conclude about the correlations between the locations of the two marbles?[5]

Exercise: You repeat the same experiment as before and you find the following results:

$$Pr(a) = \frac{90}{100}, \quad Pr(b) = \frac{10}{100}, \quad Pr(c) = \frac{0}{100}, \quad Pr(d) = \frac{0}{100}.$$

What do you suspect is the reason for such a result?[6]

Now that we have a sense of what a correlation between events is, we can define the correlation between any two events A and B as follows: If two events A and B are *independent*, then we know that the probability of the event {*A and B*} is equal to the product of $Pr(A)$ and $Pr(B)$, i.e.

$$Pr(A \cap B) = Pr(A) \times Pr(B).$$

When the two events do not fulfill this product rule, we measure the extent of dependence by the extent of deviation from the product rule, i.e. we define the correlation between the two events by the difference:

$$COR(A, B) = Pr(A \cap B) - Pr(A) \times Pr(B).$$

When this difference is positive, we say that there is *positive correlation*, and when it is negative, we say that there is *negative correlation* between events A and B.

In many problems in physics, it is more convenient to define another measure of correlation between the two events, which is as follows:

$$g(A, B) = \frac{Pr(A \cap B)}{Pr(A) \times Pr(B)}.$$

Note that this definition presumes that neither $Pr(A)$ nor $Pr(B)$ is zero.

The two correlations are related to each other. If $COR(A, B)$ is positive, this means that $g(A, B)$ is larger than 1, and vice versa. If $g(A, B)$ is negative, then $g(A, B)$ will be smaller than 1, and if the two events are independent, then $COR(A, B) = 0$, or equivalently, $g(A, B) = 1$.

Can you write down the formula connecting $COR(A, B)$ with $g(A, B)$?[7]

4.2 Conditional Probability

The concept of *conditional probability* is central to the theory of probability, perhaps even more central than the concept of probability itself. You will see why in a moment. But before defining the concept of conditional probability, let us warm up with some simple examples. This will lead us not only to the motivation for defining the concept of conditional probability, but it will also point the way towards the formal definition of this concept.

Define the event A as follows: A = {a dice shows the face with 4 dots upwards}.

Answer the following questions:

(1) What is the probability of event A?
(2) What is the probability of event A given that the dice is fair?
(3) What is the probability of event A given that the dice is fair and that we threw it in such a way that it rolled and spun in the air many times before it landed on the floor?
(4) What is the probability of event A if you know that the dice is connected to a very short rubber band or a spring that pulls the dice so that the one-dotted side will always face downwards (Fig. 4.11)?

Fig. 4.11 A dice connected to a rubber band.

(5) What is the probability of event A given that a heavy metal was attached to the face of the dice with six-dots (the one-dotted face is always parallel to the six-dotted face) (Fig. 4.12)?[8]

Fig. 4.12   A dice with a heavy metal attached to the six-dotted face.

The moral of this exercise is simple. The probability of any event is always a *conditional* probability. The condition is the given information. In most cases, the only information that we give is that the experiment is performed. We can write it symbolically as $Pr(A|\Omega)$, i.e. the probability that event A occurs given that a well-defined experiment is performed and that one of the outcomes has occurred. We usually omit the condition Ω when we ask about the probability of an event A. However, one should be aware of the fact that there exists no such thing as an *absolute* probability of an event, without giving *any* additional information. We have also mentioned this fact in Session 3, although we did not have the concept of conditional probability at that time.

Not realizing the necessity of giving additional information (i.e. that an experiment is performed, that the dice is fair, how it was thrown, etc.) can lead one to some absurd results. The famous Bertrand paradox is an example of such absurd results. Before we discuss the Bertrand paradox, let us consider some simple examples:

(1) Suppose that a dice is threaded with a string that crosses through the centers of the faces "2" and "5" (Fig. 4.13a). I tell you that I "threw" the dice by giving it a spin. What is the probability of getting a "4" i.e. the face showing upwards, perpendicular to the axis of the spring, is a "4"? The correct answer is 1/4, because there are four possible outcomes, $\{1, 3, 4, 6\}$, and we assume that these are equally probable; the outcomes "2" and "5" have probability zero.

(2) Suppose I spin the dice as in Fig. 4.13b, i.e. like a top, where three faces, say 1, 2 and 3, facing downwards, and the other faces (4, 5 and 6) facing upwards. What is the probability of obtaining the outcome "6"? Well, the answer depends on how hard I spin the dice. On the one hand, if I spun it

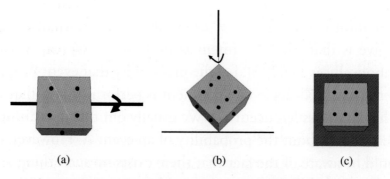

Fig. 4.13   Three experiments with "biased" dice.

gently, the chance of getting "6" is about 1/3, since there are only three possible outcomes with high probabilities. On the other hand, if I spun it harder, I might also get some of the numbers that were originally facing downwards (i.e. 1, 2 or 3). Therefore, the answer would be that the probability of obtaining a "6" is somewhat smaller than 1/3, but probably higher than the probability of obtaining a "2". If the spins were extremely hard so that I could assure you that the dice had tumbled around many times such that each outcome had a "fair" chance, the answer to my question should be 1/6.

(3) Now for the most extreme case. Suppose the dice rests on a very viscous and sticky fluid so that it can rotate about a vertical axis going through the faces "1" and "6," and initially the dice shows the face "6" pointing upwards (Fig. 4.13c). I tell you that I gave the dice a spin, and it responds by swirling around while stuck to the fluid. What is the probability of the outcome "4"? The answer is zero. So is the answer for any other number except for "6," which has a probability of one.

To conclude, in each of the examples (a), (b) and (c) we "threw" a dice. However, if we do not specify the conditions under which we carried out the experiment, we will get different answers for the same question.

Now, let us discuss a more subtle case known as the Bertrand paradox.

4.3 The Bertrand Paradox

This is a classic example demonstrating the importance of specifying exactly how an experiment is to be carried out. It is similar to the dice example discussed above. It is referred to as a paradox although it is not a paradox at all. As in the examples of Fig. 4.13, here we also have three different "solutions." The Bertrand paradox is deemed to be a paradox only because it is less obvious that the three different solutions pertain to three *different problems*, or three different experiments.

4.3.1 *The problem*

You are given a circle of 1 unit radius (Fig. 4.14). You throw a rod of length larger than the diameter of the circle (say, of length 2.2). The rod must fall on the circle in such a way that it crosses two points on the circle (Fig. 4.14). What is the probability that the length of the rod within the circle is larger than the side of an inscribed equilateral triangle? See rod A in Fig. 4.14. The following are three *different* solutions to this problem.

4.3.1.1 *First solution*

Without loss of generality of the solution, we can choose *any* point at random on the circle (we can repeat the same procedure for *any* other point — the result will be the same). From the chosen point, we draw an equilateral triangle as in Fig. 4.15a.

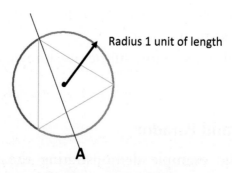

Fig. 4.14 The problem.

Choose this point as the *first* point that the rod intersects the circle. The *second* point of intersection can be any other point on the circle. Clearly, each chord that falls within the inscribed triangle has a length that is larger than the side of the triangle, and all of the others have lengths that are smaller. The three vertices of the triangle divide the total length of the circle into three equal parts. Therefore, the chords with lengths that are larger than the side of the triangle must end up in only one third of the points on the circle (see the red arc in Fig. 4.15a). Hence, the probability of having the length of the rod larger than the side of the triangle is the ratio of the length of the red arc to that of the length of the entire circle. Hence, the solution is: Probability = 1/3.

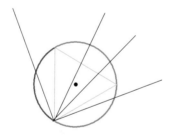

Fig. 4.15(a)   First solution.

4.3.1.2 *Second solution*

Choose one vertex of the triangle and draw the line perpendicular to the side of the triangle opposing that vertex (line *AB* in Fig. 4.15b). Draw the two points *x* and *y*, which are at distance of 1/2 from the center of the circle (*x* is on the side of the triangle opposite to A).

    Without loss of generality of the solution, we can throw the rod at random but always perpendicular to the line *AB* (we can repeat the same procedure for any other orientation of the triangle by rotating the circle). Clearly, all of the rods that fall and intersect the line *AB* between the points *x* and *y* will have a chord length larger than the side of the triangle. Since the radius of the circle is 1, and the length of the segment *xy* is also 1, the probability of the rod falling on the segment *xy* is 1/2. The solution is therefore: Probability = 1/2.

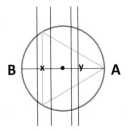

Fig. 4.15(b)   Second solution.

4.3.1.3 *Third solution*

Choose a point within the circle at random. Throw the rod in such a way that the chosen point will be the mid-point of the chord formed by the rod (Fig. 4.15c). Clearly, all of the points within the concentric circle of radius 1/2 will have a chord length longer than the side of the triangle. Since the area of the circle of radius 1/2 is a quarter of the area of the larger circle (of radius 1), the probability of choosing a point within the smaller circle is 1/4. Hence, the solution to the problem is: Probability = 1/4.

At first glance, it seems that there is a real paradox here. How can we obtain three different solutions to the same problem? You might try to look for an error in one of the solutions, but there are none. You can actually simulate the three experiments and you will get these three results. The three solutions are correct. They are simply three solutions to three different problems, i.e. the three probabilities correspond to three different experiments.

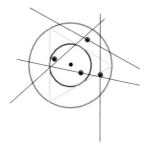

Fig. 4.15(c) Third solution.

In the case of the dice discussed in connection with Fig. 4.13, it was obvious that the three cases were different. Even though we said that we threw the dice at "random," it still left room for different *methods* of random throws. The same argument applies here — the different solutions apply to *different* random throws of the rod, or different ways of generating the random chords. In the first solution, we chose a second point of intersection at a random point on the circle; in the second solution, we chose at random point on the line *AB*; and in the third solution, we selected a random point to be the mid-point of the chord. These three procedures result in different probabilities of the outcome of the size of the chord being larger than the side of the triangle.

Thus, as in the case of the three experiments in Fig. 4.15, the *event*, the probability of which we are interested in, is the same in the three cases, but the random procedures to generate this event are different.

4.4 The General Definition of Conditional Probability

Now that we know that a probability is always a conditional probability and we already know the standard conditions (i.e.

that the dice is fair, an experiment is performed, etc.), we shall always omit this information from our notation, and refer to $Pr(A)$ as the probability of event A, presuming that the standard information is given implicitly. The next step is to calculate the conditional probability of an event A given some *new* (non-standard) information.

I throw a dice and I tell you that the result was "even." Now I ask you, "What is the conditional probability that the outcome is "4" given that the result is "even"?" We write this as:

$$Pr(\text{"4"}|even)$$

where the condition is written on the right-hand side of the vertical bar (note again the omitted standard information).

How do we calculate this conditional probability? The answer is very simple. The sample space for the dice is $\Omega = \{1, 2, 3, 4, 5, 6\}$, and the probability of each outcome is 1/6. We write this as $Pr(4) = 1/6$. Having the additional information "even" effectively *reduces* the sample space from Ω to the new sample space $\{2, 4, 6\}$. Each of these outcomes in the reduced sample space has probability 1/3, i.e. $Pr(4|even) = 1/3$.

We also notice that this result can be obtained by the ratio:

(a)
$$Pr(\text{"4"}|even) = \frac{Pr(\text{"4"})}{Pr(even)}$$
$$= (1/6)/(1/2) = 1/3.$$

This method of calculation of the conditional probability is only suggestive. Whenever we give additional information, we *reduce* the sample space from which we select the required event. But what if we are told that the result was "odd" and are asked for the probability that the outcome is "4" given that the result is "odd"? Clearly, in this case the answer is zero. But if we write the ratio as we did above, we would get the incorrect result:

(b) $$Pr(``4"|``odd") = \frac{Pr(``4")}{Pr(``odd")} = \frac{\frac{1}{6}}{\frac{1}{2}} = \frac{1}{3}.$$

This absurd result is again only indicative. The correct procedure to calculate the conditional probability is:

(c) $$Pr(``4"|even) = \frac{Pr(``4" \text{ and } ``even")}{Pr(``even")} = \frac{\frac{1}{6}}{\frac{1}{2}} = \frac{1}{3}.$$

(d) $$Pr(``4"|``odd") = \frac{Pr(``4" \text{ and } ``odd")}{Pr(``odd")} = \frac{0}{\frac{1}{2}} = 0$$

In the numerator of the first quotient (c), $Pr(``4" \text{ and } ``even")$ is simply 1/6. The numerator in (d) is $Pr(``4" \text{ and } ``odd")$, which is obviously zero. The outcome cannot be "4" and "odd" at the same time.

Before we present the general definition of conditional probability, let us work through another example that will provide us with an additional rationale for the general definition. It is also a good example for training you to think probabilistically.

We throw two fair dice at a large distance from each other so that the outcomes on each of the dice are independent. What is the probability that *both of the outcomes* are "even," given that the *sum* of the outcomes is 8?

The sample space of this experiment contains 36 elementary outcomes, which are:

Table 4.2

(1, 1), (1, 2), (1, 3), (1, 4), (1, 5), (1, 6)

(2, 1), (2, 2), (2, 3), (2, 4), (2, 5), (2, 6)

(3, 1), (3, 2), (3, 3), (3, 4), (3, 5), (3, 6)

(4, 1), (4, 2), (4, 3), (4, 4), (4, 5), (4, 6)

(5, 1), (5, 2), (5, 3), (5, 4), (5, 5), (5, 6)

(6, 1), (6, 2), (6, 3), (6, 4), (6, 5), (6, 6)

Note that the result $(1, 6)$ is different from the result $(6, 1)$. The two dice are different and distinguishable. The first number in the parenthesis is the outcome on the first dice, and the second number is the outcome on the second dice.

Assuming that all of these 36 elementary outcomes are equally likely, each has the probability of 1/36.

By inspection of Table 4.2, we can write the following probabilities:

(i) $$Pr(sum = 8) = \frac{5}{36}.$$

There are five outcomes with sum $= 8$ out of 36 total outcomes (see encircled pairs in Table 4.2). Therefore, based on the *classical* definition, we arrive at the probability 5/36.

(ii) $$Pr(sum = 8 \text{ and both outcomes are even}) = \frac{3}{36}.$$

See encircled and shaded grey pairs in Table 4.2. There are only three pairs that include two even numbers *and* sum $= 8$.

Finally, the required conditional probability is:

(iii) $$Pr(both\ numbers\ are\ even|sum = 8) = \frac{3}{5}.$$

Given the condition sum = 8, this *reduces* our sample space from 36 outcomes to only five outcomes (those encircled in Table 4.2). Out of these five outcomes, only three outcomes consist of two even numbers (encircled and shaded grey pairs in Table 4.2).

Note that we derived the last result by applying the classical "definition" of probability in the reduced sample space. We did not use any general formula. Let us see what we get if we apply methods (b) and (c) in order to calculate the conditional probability. Applying method (b), we get:

(b) $$Pr(both\ even|sum = 8) = \frac{Pr(both\ even)}{Pr(sum = 8)} = \frac{\frac{9}{36}}{\frac{5}{36}} = \frac{9}{5}.$$

Note that there are nine outcomes with both entries "even," nine outcomes with both entries "odd" and 18 outcomes with one "odd" and one "even" entry. Obviously, this method gave the wrong answer (in fact, the number is not even a probability, as it is larger than unity). Let us try the second method (c):

(c) $$Pr(both\ even|sum = 8) = \frac{Pr(both\ even\ and\ sum = 8)}{Pr(sum = 8)}$$

$$= \frac{\frac{3}{36}}{\frac{5}{36}} = \frac{3}{5}.$$

Using this method, we get the correct answer, which is the same as the one we calculated using the "classical definition." Being encouraged by calculating the conditional probability using method (c), we can generalize the definition, as well as the method of calculating the *conditional probability* of event A, *given* that event B occurred, as:

$$Pr(A|B) = \frac{Pr(A \text{ and } B)}{Pr(B)}.$$

This is a very important formula. We have obtained it by examining a few simple examples. (I encourage you to invent other examples and see that this formula gives the correct result. If you cannot invent an example, let me suggest one: Calculate the probability that the two outcomes on the two dice are "even" given that the sum of the outcomes is "4.")

We shall soon see how the conditional probability is related to the correlation between the two events. Before doing this, keep in mind that there is always the hidden condition Ω that we omit for simplicity of the notation. Furthermore, keep in mind that the formula given above is meaningful only if event B is *not* an impossible event, i.e. that $Pr(B) \neq 0$. It is meaningless to say that an event that has a zero probability has occurred.

4.5 Correlation Between Events and Conditional Probability

We can now connect the notion of correlation and the conditional probability. In doing so, we obtain a different view of the meaning of both correlation and conditional probability. We also obtain an additional rationale for the formula for the conditional probability.

In this section, we shall use the definition of correlation as the ratio of the probability of the joint event $A \cap B$ and the product of the probabilities of each of the events A and

B, i.e.:

$$g(A, B) = \frac{Pr(A \cap B)}{Pr(A)Pr(B)}.$$

Recall that this quantity measures the extent of dependence between the two events A and B. When the two events are independent, then $g(A, B) = 1$. When they are dependent, we distinguish between the two cases. If $g(A, B) > 1$, we say that the correlation is *positive*. The larger the value of $g(A, B)$, the larger the extent of the (positive) dependence. On the other hand, if $g(A, B) < 1$, then we say that the correlation is *negative*; the smaller the value of $g(A, B)$, the more (negative) the dependence between the two events. Note that the limit of negative dependence is when $g(A, B) = 0$; this is equivalent to the impossibility of the event $A \cap B$, i.e. the two events are mutually exclusive. They cannot occur together. Note also that $g(A, B)$ is defined *symmetrically* with respect to A and B. This means that $g(A, B)$ is the same as $g(B, A)$.

Now we combine the conditional probability $Pr(A|B)$ with correlation $g(A, B)$:

$$Pr(A|B) = \frac{Pr(A \cap B)}{Pr(B)} = \frac{Pr(A)Pr(B)g(A, B)}{Pr(B)} = Pr(A)g(A, B).$$

Examine carefully the meaning of this relationship between $Pr(A)$ and $Pr(A|B)$.

Recall that $Pr(A)$ is the unconditional probability of A (remember the condition Ω is always there but omitted). The last equation shows how the unconditional probability $Pr(A)$ is *modified* when some additional information (B) is given.

Another way of rewriting the same equation is:

$$g(A, B) = \frac{Pr(A|B)}{Pr(A)}.$$

When the two events A and B are independent, then $g(A, B) = 1$, and the probability of A is not modified by giving the new information B. On the other hand, when $g(A, B) > 1$, i.e. when the conditional probability $Pr(A|B)$ is larger than the original probability $Pr(A)$, we can say that event B "supports" or "enhances" the probability of event A. Thus, positive correlation is equivalent to positive difference:

$$Pr(A|B) - Pr(A) > 0, \quad (\text{when } g(A, B) > 1).$$

When $g(A, B) < 1$, the correlation is negative. The additional information B *reduces* the probability of A. We can say that B does not support A, or B *diminishes* the probability of A. Equivalently, negative correlation is the same as negative difference:

$$Pr(A|B) - Pr(A) < 0, \quad (\text{when } g(A, B) < 1).$$

Recall that $g(A, B)$ is defined symmetrically with respect to A and B. On the other hand, the conditional probability is not symmetric with respect to A and B. In other words, $Pr(A|B)$ is not the same as $Pr(B|A)$ (provided that these are definable, i.e. $Pr(B) \neq 0$ and $Pr(A) \neq 0$). In other words, the probability of A knowing B is, in general, different from the probability of B knowing A. However, the *sign* of the *correlation* is the same for both of the conditional probabilities, i.e. if A supports B, then B supports A, and if A does not support B, then B does not support A. We write this is as:

$$\frac{Pr(A|B)}{Pr(A)} = \frac{Pr(B|A)}{Pr(B)} = g(A, B).$$

When $g(A, B) = 1$, both ratios are equal to one, the information B does not affect the probability of A and information A does not affect the probability of B. When $g(A, B) > 1$, both A supports B and B supports A. When $g(A, B) < 1$, A does not support B and B does not support A.

Next, we offer another perspective of the conditional probability. For two independent events A and B, we write:

$$Pr(A \cap B) = Pr(A) \times Pr(B).$$

When the two events are dependent, we "correct" this formula by *multiplying* by the correlation function:

$$Pr(A \cap B) = Pr(A) \times Pr(B) \times g(A, B)$$
$$= Pr(A) \times Pr(B|A).$$

On the right-hand side of this equation, we see a product of two probabilities. This product suggests an interpretation of the probability of the intersection as a product of the probabilities of two *independent* events, in this case A and (B|A). Thus, although we started with two *dependent* events A and B, the two *new events* A and (B|A) behave as if they were *independent*.

Recall that $Pr(A \cap B)$ is symmetric with respect to A and B. We can write it in two different ways:

$$Pr(A)Pr(B|A) = Pr(A \cap B) = Pr(B)Pr(A|B).$$

On the right-hand side of this equation, we have the probability that B occurred and the probability that (A|B) occurred. On the left-hand side, we have the probability that A occurred and the probability that (B|A) occurred. At the center of this equation, we have the probability that both A *and* B occurred.

The following example should clarify this view. We throw a dice and ask for the probability of the event: $A \cap B$ = ("6" and "even"). The two events "6" and "even" are dependent, but we can also write the probability of the event $A \cap B$ as:

$$Pr(\text{"6" and "even"}) = Pr(6) \times Pr(\text{"even"}|6) = \frac{1}{6} \times 1 = \frac{1}{6}$$

$$= Pr(\text{"even"}) \times Pr(\text{"6"}|\text{"even"})$$

$$= \frac{1}{2} \times \frac{1}{3} = \frac{1}{6}.$$

The event "6" and the event "even" are dependent, given that one of these affects the probability of the other event. However, the two events "6" and ("even"|"6") are independent. Knowing "6" is already taken into account in the event ("even"|"6"), and therefore makes the latter independent of the event "6." Likewise, the event "even" is independent of the event ("6"|"even"), and in general, A and (B|A) are independent since the condition A in the second already takes into account the dependence between the two events A and B; therefore, the events A and (B|A) are independent.

We present here an experimental example where one can "see" and even *measure* the correlation between two events.

Suppose we have a surface on which there are two sites *a* and *b* (Fig. 4.16). In biological systems, there are many molecules that *bind* to proteins and to DNA. One of the most important examples is the hemoglobin molecule, which has four sites for binding oxygen molecules (Fig. 4.17). Here, we shall discuss a simpler example. We have two sites (a) and (b) and a molecule *s* that can bind to either or both of these sites.

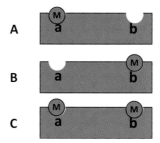

Fig. 4.16 A surface with two sites on which a molecule M can bind.

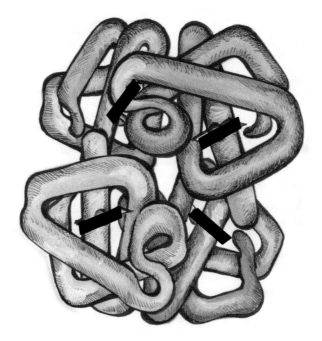

Fig. 4.17 Schematic structure of Hemoglobin, with four binding sites for oxygen, shown as black bars.

Define the following events:

$$A = \{\textit{site a is occupied}\}$$
$$B = \{\textit{site b is occupied}\}$$
$$A \cap B = \{\textit{site a and site b are occupied}\}.$$

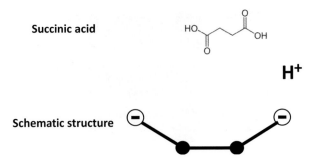

Fig. 4.18 A succinic acid is shown as a molecule with two binding sites for hydrogen ion.

Suppose that the probabilities of these three events (Fig. 4.16) were measured and the results were found to be:

$$Pr(A) = \frac{1}{12}, \quad Pr(B) = \frac{1}{10}, \quad Pr(A \cap B) = \frac{1}{200}.$$

What can you conclude from these measurements?[9]

Here is a real example, for which these probabilities are measurable.

Dicarboxylic acid (Fig. 4.18) may be viewed as a molecule with two binding sites for hydrogen ions (protons), denoted H^+. The negative correlation between the two events A and B is, in this case, a result of the (Coulombic) repulsion between the two positively charged protons.

Exercise: You are given two molecules, each having two carboxylic acid groups, as shown in Fig. 4.19. You do not need to know what a carboxylic acid is. Simply view the two structures in the figure as molecules that can bind one or two hydrogen ions, or protons (H^+) at the locations indicated by the minus signs in Fig. 4.19. In which of the two molecules in Fig. 4.19 is the correlation between the two protons more negative? What do you expect to be the correlation between the two protons when the two sites are very far apart?[10]

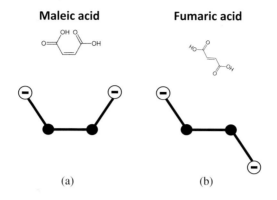

Fig. 4.19 (a) Fumaric and (b) Maleic acids.

In the case of the two protons (H^+) binding to negatively charged sites, it is relatively easy to understand the origin of the negative correlation. This is simply a result of the repulsion between the two charges of the same sign.

I cannot resist the temptation to tell you the story of the binding of oxygen to hemoglobin. Hemoglobin is a large protein with four binding sites, each binding one oxygen molecule (O_2). This is an extremely important molecule. It is the molecule that transports oxygen from the lungs to all of the other parts of the body where oxygen is needed. It was found experimentally that the probability of binding an oxygen molecule to a second site, given that the first site is occupied, is *larger* than the probability of binding to the first site. We write this as:

$$Pr(2|1) > Pr(2) = Pr(1).$$

Here, binding to the "first site" means binding to any site (it could be site "1," "2," "3" or "4") provided all other sites are empty. This means that there is a *positive* correlation between the two oxygen molecules binding to two sites. For a long time, the origin of this positive correlation was a mystery. The two sites are very far apart and there is no obvious attraction between the two oxygen molecules. In fact, it was found that there is also a positive correlation for binding on the third and the fourth sites. This phenomenon is called the *cooperativity* of the binding of oxygen to hemoglobin. I will not bother you with any further details on this extraordinary phenomenon that is responsible for the *efficiency* of the transportation of oxygen by hemoglobin.[11]

4.6 Dependence and the Extent of Overlapping Between Two or More Events

We shall discuss here two examples, one discrete and one continuous, in order to demonstrate the relationship between the extent of overlapping events and the extent of correlation between events.

Let us consider the following case. In a given roulette, there are 12 numbers altogether: $\{1, 2, 3, 4, 5, 6, 7, 8, 9, 10, 11, 12\}$.

Each of us chooses a sequence of six consecutive numbers, say I choose the sequence:

$$A = \{1, 2, 3, 4, 5, 6\}.$$

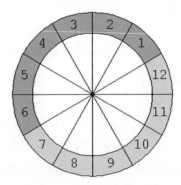

Fig. 4.20 A roulette with 12 outcomes, with two regions A (blue) and B (yellow).

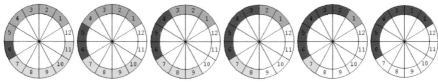

Fig. 4.21   Different choices of the event B. The intersection of B with A is shown in green.

Call "A" *my* "territory." You choose the sequence:

$$B = \{7, 8, 9, 10, 11, 12\}.$$

Call "B" *your* "territory."

The ball is rolled around the ring of the roulette. We assume that the roulette is "fair," i.e. each outcome has the same probability of 1/12. If the ball stops at my "territory," i.e. if it stops at any of the numbers I chose $\{1, 2, 3, 4, 5, 6\}$, then I win. If the ball stops in your "territory," i.e. if it stops at any of the numbers you chose $\{7, 8, 9, 10, 11, 12\}$, then you win.

Clearly, each of us has a probability 1/2 of winning. The ball has equal probability of 1/12 of landing in any number, and each of us has six numbers in each "territory." Hence, each of us has the same chances of winning.

Now, suppose we run this game and you are told that I won. What is the probability that you will win if you chose B? Clearly, $Pr(B|A) = 0 < \frac{1}{2}$, i.e. the conditional probability of B given A is zero, which is *smaller* than the unconditional probability $Pr(B) = \frac{1}{2}$.

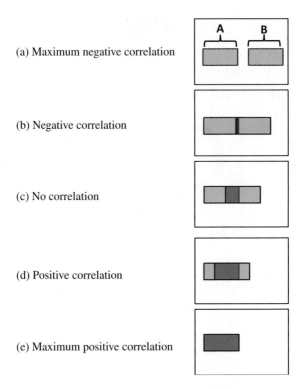

(a) Maximum negative correlation

(b) Negative correlation

(c) No correlation

(d) Positive correlation

(e) Maximum positive correlation

Fig. 4.22 Different extent of overlapping between A and B.

Exercise: Calculate the following conditional probabilities. In each example, my choice of the sequence $A = \{1, \ldots, 6\}$ is *fixed*. Calculate the conditional probabilities for the following *different* choices of your sequence:

$$Pr(7, 8, 9, 10, 11, 12|A), \quad Pr(6, 7, 8, 9, 10, 11|A),$$
$$Pr(5, 6, 7, 8, 9, 10|A), \quad Pr(4, 5, 6, 7, 8, 9|A)$$
$$Pr(3, 4, 5, 6, 7, 8|A), \quad Pr(2, 3, 4, 5, 6, 7|A),$$
$$Pr(1, 2, 3, 4, 5, 6|A)^{12}.$$

The second example is shown in Fig. 4.22. Consider a square board, on which we draw two rectangles A and B having the same dimensions $a \times b$. A dart can hit any point on the board. The two rectangles are part of the total board.

Since A and B have the same area, the probability of hitting A is equal to the probability of hitting B. Let us denote that by $Pr(A) = Pr(B) = p$, where p is the ratio of the area of A to the area of the total board; say $p = 1/10$ in this illustration. This means that the area of either A or B is 1/10 of the total area of the board.

When the two rectangles are separated, i.e. there are no overlapping area, we have:

$$Pr(A|B) = 0.$$

This is the case of an extreme *negative* correlation. Given that the dart hits B, the probability that it hits A is zero (i.e. smaller than p). This result is the same as the first case in the example with the roulette.

The other extreme case is when A and B become congruent, i.e. the overlapping area is total. In this case, we have:

$$Pr(A|B) = 1.$$

This is the extreme *positive* correlation. Given that the dart hits B, the probability that it hits A is one (i.e. larger than p).

When we move B towards A, the correlation changes continuously from maximal *negative* correlation (non-overlapping) to *maximal* positive correlation (total overlapping). In between, there is a point when the two events A and B are independent. Can you calculate this point?[13]

Instead of rectangles, you can take any two regions, e.g. circles, ellipses, etc. When they are separated (no overlap), then they are negatively correlated. When they are brought closer to each other so that they overlap, the correlation changes from negative to positive (Fig. 4.23).

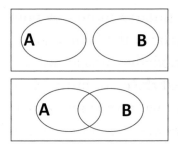

Fig. 4.23 Two ellipses with no overlap and with overlap.

If you like geometry, try to calculate the distance between the two circles in which there is no correlation. This is not an easy problem. It requires some knowledge of the geometry of circles and sections of a circle.

Note carefully that neither A nor B are the certain event (i.e. the entire board).

4.7 Conditional Probability and Subjective Probability

There is a tendency to refer to "probability" as *objective*, and to *conditional* probability as *subjective*. First, recall that probability is always *conditional*. When we say that the probability of the outcome "4" of throwing a dice is 1/6, we actually mean that the *conditional* probability of the outcome 4 *given* that one of the possible outcomes $\{1, 2, 3, 4, 5, 6\}$ has occurred or will occur, that the dice is honest and that we threw it at random, as well as any other information that is relevant. We usually suppress this *given* information in our notation and refer to it as the unconditional probability. This is considered to be an *objective* probability.

Now let us consider the following two pairs of examples: ("O" stands for objective and "S" stands for subjective):

O_1: The conditional probability of an outcome "4," given that Jacob *knows* that the outcome is "even," is 1/3.

O_2: The conditional probability of an outcome "4," given that Abraham *knows* that the outcome is "odd," is zero.

S_1: The conditional probability that the "defendant is guilty," given that he was seen by the police at the scene of the crime, is 90%.

S_2: The conditional probability that the "defendant is guilty," given that he was seen by at least five persons in another city at the time of the crime's commission, is nearly 0%.

In all of the aforementioned examples, there is a tendency (such statements are sometimes made in textbooks) to refer to the *conditional* probability as a *subjective* probability. The reason is that, in all the aforementioned examples, we included *personal knowledge* of the conditions. Therefore, we judge that it is highly *subjective*. However, that is not so. The two probabilities denoted O_1 and O_2 are objective probabilities. The fact that we mention the names of the persons who are knowledgeable of the conditions does not make the conditional probability subjective. We could make the same statement as in O_2, but with Rachel instead of Jacob: The conditional probability of an outcome "4," given that Rachel knows that the outcome is even, is 1/3. The result is the same. The apparent subjectivity of this statement is a result of the involvement of the *name* of the person who "knows" the condition. A better way of rephrasing O_2 is: The conditional probability of an outcome "4," given that *we* know that the outcome is even, is 1/3; or even better: The conditional probability of an outcome "4," given that the outcome is "even," is 1/3.

In the last two statements, it is clear that the fact that Jacob, Rachel or anyone of us *knows* the condition does not have any effect on the conditional probability. In the last statement, we made the condition completely impersonal. Thus, we can conclude that the *given condition* does not, in itself, convert an objective (unconditional) probability into a subjective probability.

To the best of my knowledge, the probabilities used in all cases in the sciences are objective. The reason is that the knowledge of the probabilities is usually explicitly or implicitly *given*. There is a general agreement that there are essentially two distinct types of probabilities. One is referred to as the judgmental probability, which is highly subjective — the two statements S_1 and S_2 above fall into this category. The second type are physical or scientific probabilities, which are considered to be objective probabilities.

In using probability in the sciences, we always presume that the probabilities are given either explicitly or implicitly by a given formula of how to calculate these probabilities. Sometimes, these are very easy to calculate, and sometimes extremely difficult, but you always assume that they are "there" in the event, as much as mass is attached to any piece of matter.

4.8 Conditional Probability and Cause and Effect

The "condition" in the conditional probability of an event may or may not be the *cause* of the event. Consider the following two examples:

(1) The conditional probability that the patient will die of lung cancer, given that he or she has been smoking for many years, is 0.9.

(2) The conditional probability that the patient is a heavy smoker, given that he or she has lung cancer, is 0.9.

Clearly, the information given in the first case is the *cause* (or the very probable cause) of the occurrence of lung cancer. In the second example, the information that is given in the condition — that the patient has cancer — certainly cannot be the *cause* for being a heavy smoker. The patient could have started to smoke at age 20, far prior to the time that the cancer developed.

Although the two examples given above are clear, there are cases where conditional probability is confused with causality. Since we perceive causes as preceding their effects on the time axis, so it is also with the "condition" in conditional probability; it is conceived as occurring earlier in the time axis.

Consider the following simple and very illustrative example that was studied in great detail by Falk (1979).[14]

You can view it as a simple exercise in calculating conditional probabilities. However, I believe this example has more to it. It demonstrates how we intuitively associate conditional probabilities with the arrow of time, confusing causality with a conditional probabilistic argument.

The problem is very simple: An urn contains four balls — two white balls and two black balls. The balls are well mixed and we draw one ball blindfolded.

First we ask: What is the probability of the event "a white ball is drawn on the first draw"? The answer is immediately clear: 1/2. There are four equally probable outcomes; two of them are consistent with the "white ball" event, and hence, the probability of the event is $\frac{2}{4} = \frac{1}{2}$.

Next we ask: What is the conditional probability of drawing a white ball on a *second* draw, given that in the first draw, we drew a white ball (the first ball is not returned to the urn)? We write this conditional probability as $Pr(White_2|White_1)$. The calculation is very simple. We know that a white ball was drawn on the first trial and was not returned. After the first draw, there are three balls left — two blacks and one white. The probability of drawing a white ball is simply 1/3.

This is quite straightforward. We write:

$$Pr(White_2|White_1) = 1/3.$$

Now, for the trickier question: What is the probability that we *drew* a white ball in the *first* draw, given that the *second* draw was a white ball? Symbolically, we ask for:

$$Pr(White_1|White_2) = ?$$

This is a baffling question. How can an event in the "present" (white ball drawn in the *second* trial) affect an event in the "past" (white ball drawn in the *first* trial)?

These questions were actually asked in a classroom. The students easily and effortlessly answered the question about $Pr(White_2|White_1)$, arguing that drawing a white ball in the first draw has *caused* a change in the urn, and therefore has influenced the probability of drawing a second white ball.

However, asking about $Pr(White_1|White_2)$ caused uproar in the class. Some claimed that this question is meaningless, arguing that an event in the present cannot affect the probability of an event in the past. Some argued that since the event in the present cannot affect the probability of the event in the past, the answer to the question is 1/2. They were wrong. The answer is 1/3. We shall leave this question for a while. We shall revert to this problem and its solution after presenting Bayes' theorem at the end of the next session (see solution to the urn problem at the end of Session 5).

The distinction between causation and conditional probability is important. Perhaps we should add one characteristic property of causality that is not shared by conditional probability: Causality is transitive. This means that if A causes B, and B causes C, then A causes C; symbolically: If $A \Rightarrow B$ and $B \Rightarrow C$, then $A \Rightarrow C$. A simple example: If smoking causes cancer, and cancer causes death, then smoking causes death.

Conditional probability might or might not be transitive. We have already distinguished positive correlation (or supportive correlation) from negative correlation (or counter- or anti-supportive correlation).

If A supports B, i.e. the probability of the occurrence of B *given A*, is larger than the probability of the occurrence of B i.e. $Pr(B|A) > Pr(B)$, and if B supports C, i.e. $Pr(C|B) > Pr(C)$, then in general, it does not follow that A supports C.

Here is an example where supportive conditional probability is not transitive. Consider the following three events in throwing a dice:

$$A = (1, 2, 3, 4), \quad B = (2, 3, 4, 5), \quad C = (3, 4, 5, 6)$$

Clearly, A supports B, i.e. $Pr(B|A) = \frac{3}{4} > Pr(B) = \frac{2}{3}$ and B supports C, i.e. $Pr(B|A) = \frac{3}{4} > Pr(B) = \frac{2}{3}$, but A does not support C, i.e. $Pr(C|A) = \frac{1}{2} < Pr(C) = \frac{2}{3}$.

Another example is shown in Fig. 4.24 in terms of a Venn diagram. Here also, A supports B and B supports C, but A does not support C.[15]

Here is another surprising and perhaps counter-intuitive example of conditional probabilities. Suppose the evidence E_1 supports some conclusion C. In addition, some other evidence

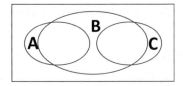

Fig. 4.24 A supports B, B supports C, but A does not support C.

E_2 supports the same conclusion C. Do you think that *both* evidence E_1 *and* evidence E_1 support or do not support the conclusion C?

While pondering on the answer to this question, enjoy the following story.

4.8.1 *Can more incriminating evidence be inculpatory?*

Mary and Bob Victimsberg lived in a wooded area in the suburbs. On New Year's Eve, the couple was invited to a midnight celebration in the city. They left the house at 12 midnight and were back at 1 a.m. When they entered the house, they immediately knew that someone had broken in. All of their valuables were gone. The infrared detector, which is activated every time there is some movement in the house, was blinking. They checked the detector and found that for exactly 30 min, someone had been moving in the house. The reading on the infrared detector, however, did not tell them at exactly what time the movement had been detected. All that they could say was that it had happened between 12 midnight and 1 a.m., the time frame during which they were out of the house. They immediately called the police to report the break-in. They were asked if they had any suspects in mind, but they said no. The only thing they could say with certainty was that they were out of the house between 12 midnight and 1 a.m., and that the infrared detector was activated for half an hour between 12 midnight and 1 a.m.

The following day, Jerry Pick, a well-known criminal, was apprehended for shoplifting in a supermarket not far from where the Victimbergs lived. While the county policeman was writing down the case in his police blotter, the thought crossed his mind that Pick must have been involved in the Victimbergs household break-in, so he decided to detain Pick for the night. Pick denied any involvement in the break-in.

Pick's detention made it to the local papers, as well as on TV, with his face plastered on TV and across the papers. News features both on TV and in the papers mentioned that there had been an unsolved robbery the day before, and that the police had requested the public to come forward if they had seen any suspects in the vicinity of the Victimberg's house on New Year's Eve between 12 midnight and 1 a.m. It was similarly mentioned in the news that the burglary took place for exactly 30 min, as the infrared detector had indicated, although the exact time of the break-in was not known.

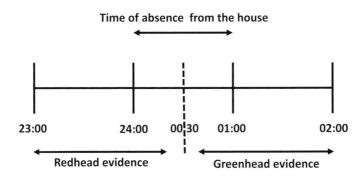

Fig. 4.25 Times in which the various events occur.

Chief Officer Greenhead was sitting in a bar, nursing a mug of beer when the TV screen flashed Pick's face. The bar was a mere 5-min walk from where the Victimbergs lived. While glued to the news about the two robberies from the past two days, it suddenly dawned on Chief Greenhead that Pick had entered the very same bar he was in on New Year's Eve at exactly 12.35 a.m. He also vividly recalled that Pick had stayed in the bar until 2 a.m. After performing a mental calculation and considering Pick's criminal record, Chief Greenhead was convinced that Pick could have committed the robbery between 12 midnight and 12.30 a.m., and had enough time (5 min) to get from the Victimberg's residence to the bar. See Fig. 4.25. With this surefire evidence, he had written down all of his recollections from New Year's Eve and sent them to the court, fully convinced that his circumstantial evidence would bolster the evidence against Pick in terms of his involvement in the Victimberg's robbery.

At exactly the same time that Chief Greenhead was rushing to complete his report at the bar, Officer Redhead was dining in a restaurant nearby. Officer Redhead was just as familiar with Pick's criminal record and, by coincidence, he had also seen Pick dining in the same restaurant where he was currently dining on the night of the alleged break-in at the Victimberg's. He vividly recalled that Pick had dined there at 11 p.m. — just before midnight — and had left the restaurant at 12.25 a.m. See Fig. 4.25. Watching the news on TV, he, like Chief Greenhead, carried out a quick mental calculation and concluded that Pick had enough time to get from the restaurant to the Victimberg's residence after he had left the restaurant, stayed and robbed the house in 30 min and then sped off. With hands trembling, he wrote down all of the details and was satisfied with his analysis. With a glimmer of hope, he felt confident that his accurate details and analysis would land him a promotion.

Almost simultaneously, the two reports were well on their way to court, with their bearers equally confident of the veracity of their details. Both believed that their reports contained highly plausible evidence that could pin Pick to involvement in the New Year's Eve break-in. Pick, of course, vehemently denied the allegations.

The next morning, the two reports were laid out on top of Judge Judy's desk. She first read Chief Greenhead's report. It was very convincing, considering Pick's criminal record. So convincing was Chief Greenhead's report that Judge Judy almost issued an extension of Pick's detention in the county jail. Upon reading Officer's Redhead report, however, which also contained very convincing evidence against Pick, Judge Judy decided to let Pick off the hook.

Can you explain why Judge Judy changed her mind?[16]

This example shows that if B supports A and C also supports A, it does not necessarily follow that (B and C) supports A. See Notes 16 and 17 and Figs. 4.26 and 4.27.

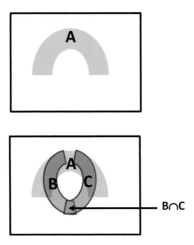

Fig. 4.26 B supports A , C supports A, but B *and* C does not support A.

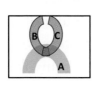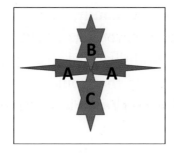

Fig. 4.27 Two examples showing: B does not support A, C does not support A, but B and C supports A.

4.8.2 *Distinction between disjoint and independent events*

It is important to distinguish between *disjoint* (i.e. mutually exclusive) and *independent* events. Disjoint events are events that are mutually exclusive; the occurrence of one excludes the occurrence of the second. The events "even" and "5" are disjoint. In terms of Venn diagrams, two regions that are non-overlapping are disjoint. If the dart hits one region, say *A*, in Fig. 4.22a, then we know for sure that it did not hit the second region *B*.

Disjoint events are properties of the events themselves (i.e. the two events have no common elementary events). Independence between events is not defined in terms of the elementary events that are contained in the two events, but is *defined* in terms of their probabilities. If the two events are disjoint, then they are strongly *dependent*. In the above example, we say that the two events are *negatively* correlated. In other words, the conditional probability of hitting one region, given that the dart hit the other region, is zero (which is smaller than the "unconditional" probability of hitting the region in question).

If the two regions A and B do overlap (i.e. they are not disjoint), then the two events could be either dependent or independent. In fact, the two events could either be positively or negatively correlated. The reader is urged to review the two examples shown in Fig. 4.22 and Fig. 4.23.

4.9 Pairwise and Triplewise Independence Between Events

We present here two examples where dependence or independence between two events does not imply dependence or independence between three events, and vice versa. Three events are said to be independent if and only if the joint probability of the three events is the product of the probabilities of the three events.

4.9.1 *Pairwise independence does not imply triplewise independence*

Consider a board of total area S being hit by a dart. The probability of hitting a certain region is assumed to be proportional to its area. On this board, we draw three regions A, B and C (Fig. 4.28). If the area of the entire board is chosen as unity, and the areas of A, B and C are 1/10 of S, then we have the

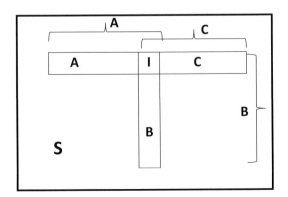

Fig. 4.28 Three events A, B and C and the intersection I.

probabilities:

$$Pr(S) = 1, \quad Pr(A) = Pr(B) = Pr(C) = \frac{1}{10}.$$

Note that the intersection I is included in A, B and C.

In this example, we have chosen the regions A, B and C in such a way that the area of the intersection (region I) is 1/100 of S. Hence, in this case we have:

$$Pr(A \text{ and } B) = Pr(B \text{ and } C) = Pr(A \text{ and } C) = \frac{1}{100}$$
$$= Pr(A)Pr(B) = Pr(A)Pr(C) = Pr(B)Pr(C).$$

Thus, we have pairwise independence, e.g. the probability of hitting A and B is the product of the probabilities $Pr(A)$ and $Pr(B)$. However, in this system:

$$Pr(A \text{ and } B \text{ and } C) = \frac{1}{100} \neq Pr(A)Pr(B)Pr(C) = \frac{1}{1000}.$$

The probability of hitting A, B and C is not equal to the product of the probabilities $Pr(A)$, $Pr(B)$ and $Pr(C)$, i.e. there is no triplewise independence.

4.9.2 Triplewise independence does not imply pairwise independence

In this example (Fig. 4.29), the total area is again S. In addition, the area of each region A, B and C is again 1/10 of S. But now the intersection I of the three regions has an area 1/1000 of S. In this case, we have:

$$\frac{1}{1000} = Pr(A \text{ and } B \text{ and } C)$$

$$= Pr(A)Pr(B)Pr(C) = \frac{1}{10} \times \frac{1}{10} \times \frac{1}{10}.$$

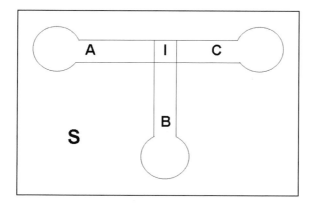

Fig. 4.29 Three events A, B and C and the intersection I.

The probability of the event (*A and B and C*) is the product of the three probabilities $Pr(A)$, $Pr(B)$ and $Pr(C)$. On the other hand:

$$\frac{1}{1000} = Pr(\text{A and B}) \neq Pr(\text{A})Pr(\text{B}) = \frac{1}{100}$$

and similarly for $Pr(A \text{ and } C)$ and $Pr(B \text{ and } C)$. Therefore, the triplewise independence does not necessarily imply pairwise independence.

Exercise: A four-faced dice is colored as follows: One face is all red (R), one face is all blue (B), one face is all green (G) and one face has all three colors R, B and G (say, 1/3 area of each color).

Show that, in this example, there is independence between pairs of colors, but not independence between triplets.[18]

4.9.3 *Confusing conditional probability and joint probability*

I had a friend who used to ride a motorcycle. One night, while driving on the highway, he was hit by a truck and was seriously injured. When I visited him in the hospital, he was beaming and in a euphoric mood. I was sure that this was because of his quick and complete recovery. To my surprise, he told me that he had just read an article in which statistics about the frequencies of car accidents were reported. It was written in the article that the chances of being involved in a car accident over a lifetime is one in a thousand. The chances of being involved in two accidents over a lifetime is about one in a million. Hence, he happily concluded: "Now that I've had this accident, I know that the chances of me being involved in another accident are very small . . . " I did not want to dampen his high spirits, but he was clearly confusing the probability of "having two accidents in a lifetime" with the conditional probability of "having a second accident, given that you were involved in one accident."

Of course, he might have been right in his conclusion that his chances of being involved in another accident is small. However, his probabilistic reasoning was wrong. If the accident was a result of his fault, then he might take steps to be very careful in the future, and might avoid driving on the highways, at night or stop riding a motorcycle altogether. All of these actions would reduce his chances of being involved in a second accident. But this argument implies that there is a *dependence* between the two events, i.e. the "given condition" affects the chances of being involved in a second accident. If, however, there is no dependence between the events, say, if the accident was not his fault, even if he was extremely careful in the future, the chances of his being involved in another accident would not be reduced merely because he was involved in a previous accident!

Let us make the arguments more precise. Suppose that you tossed a coin that was known to be fair 1000 times, and all of the outcomes turned out to be heads (H). What is the probability of a H in the next throw given that the previous 1000 throws were

H? We assume that the throws are independent. Most untrained people would say that the chances of having 1001 heads are extremely small. That is correct. The chances are $(\frac{1}{2})^{1001}$, which is extremely small indeed. But the question was on the *conditional* probability of the next outcome being H, given 1000 heads in the last 1000 tosses. This conditional probability is one half (assuming that the coin is fair and all of the events are *independent*).

The psychological reason for the confusion is that you know that the probability of H and tails (T) is half. So if you normally make 1000 throws, it is most likely that you will get about 500 H and 500 T. Given that the first 1000 throws resulting in H (although a very rare event) is possible, you might feel that "it is time that the chances will turn in favor of T" and the sequence of outcomes beginning to behave properly. Therefore, you feel that the chances of T, given that 1000 H outcomes have occurred, are now close to one. That is wrong, however. In fact, if a coin shows 1000 outcomes in a row to be H, I might suspect that the coin is unbalanced, and therefore I might conclude that the chances of the next H are larger than 1/2.

To conclude, if we are given a fair coin, and it is tossed at random (which is equivalent to saying that the probability of H is 1/2), the probability of having 1000 outcomes of H is very low, at $\left(\frac{1}{2}\right)^{1000}$, but the *conditional* probability of having the next H, given "1000 H in a row," is still 1/2 . This is of course only true assuming that the events resulting from each toss are independent.[19]

The last example is also known as the gambler's fallacy. A gambler bets on the outcome T. For 100 trials, he gets continuous H outcomes and loses. He "feels" that, next time, the odds must turn in his favor and the probability of getting a T will therefore be larger than 1/2.

4.9.3.1 *A challenging problem*

The following is a problem of significant historical value. It is considered to be one of the problems to which the solution not only crystallized the concept of probability, but also transformed the reasoning processes about chances that are followed in gambling salons into mathematical reasoning processes that occupy the minds of mathematicians. This problem was addressed to Blaise Pascal by his friend Chevalier de Mere in 1654.

Two players bet $10 each. The rules of the game are very simple. Each one chooses a number between 1 and 6. Suppose Dan chose 4 and Linda chose 6. They roll a single dice and record the sequence of the outcomes. Every time an outcome "4" appears, Dan gets a point. When a "6" appears, Linda gets a point. The player who collects three points first wins the total sum of $20. For instance, a possible sequence could be: 1, 4, 5, 6, 3, 2, 4, 4, ...

Once a "4" appears for the third time, Dan wins the entire sum of $20 and the game ends.

Now, suppose the game is started and at some moment the sequence of outcomes is the following:

$$1, 3, 4, 5, 2, 6, 2, 5, 1, 1, 5, 6, 2, 1, 5.$$

At this point, there is some emergency and the game must be stopped! The question is how to divide the sum of $20.

Note that this problem does not arise if the rules of the game explicitly instruct the players on how to divide the sum should the game be halted. But in the absence of such a rule, it is not clear how to divide the sum.

Clearly, one feels that since Dan has "collected" one point, and Linda "collected" two points, Linda should get the larger portion of the $20. But how much larger? The question is, what is the *fairest* way of splitting the sum, given this sequence of outcomes? But what does it mean by the *fairest splitting of the sum*? Should Linda get twice as much as Dan because she has gained twice as many points in her favor? Or perhaps, should they simply split the sum into two equal parts since the winner is undetermined? Or perhaps, should they let Linda collect the total sum because she is closer to winning than Dan.

A correspondence between Blaise Pascal and Pierre de Fermat ensued for several years. These were the seminal thoughts that led to the theory of probability. Recall that in the 17th century, the concept of probability was far from being crystallized. The difficulty was not only of finding the mathematical solution to the problem. It was not less difficult to clarify what the problem was, i.e. what does it mean to find a *fair* method of splitting the sum?

The answer to the last question is the following: Having no specific rule on how to divide the sum in the case of halting the game, the "fairest" way of splitting the sum is according to the *ratio* of the *probabilities* of the two players winning the game, had the game continued.

In stating the problem in terms of probabilities, one hurdle was overcome. We now have a well-formulated problem. But how do we calculate the *probabilities* of either player winning? We feel that Linda has a better chance of winning since she is "closer" to collecting three points than Dan.

We can easily calculate that the probability of Dan winning on the *next* throw is zero. The probability of Linda winning on the next throw is 1/6, and the probability of neither one of them winning on the next throw is 5/6. One can calculate the probabilities of winning after two throws, three throws, etc. The calculations become very complicated for the fourth, fifth and larger numbers of throws. Try calculating the probability of each player winning in the next two throws and for the next three throws and see just how messy it gets. There is a simple solution based on the theorem of total probability that involves a one-unknown equation.[20]

Exercise: Solve the same problem for the following two cases:

(1) Dan has collected two points and Linda has zero points.
(2) Dan has collected one point and Linda has zero points.

4.9.3.2 *Treasure hunt*

Can you help me in finding the treasure?

Not long ago, I received a letter from the King of Goldland. He wrote to tell me that he had read my book *Entropy Demystified* and that he enjoyed reading it immensely. As a token of his gratitude, he said that he had buried a treasure for me in a deserted island, called Fortune Island, which is off the coast of Kameya-meya. He enclosed a crudely drawn map, perhaps something that he himself had drawn, with the following instructions:

> When you reach the northern part of the island, you will see two tall, majestic palm trees, with green, abundant fronds swaying gracefully with the wind.

I was amused at the way the king wrote. He was not only generous with his gift to me, he was poetic as well. He went on with his instructions:

> Between those two stately palm trees, you will find wooden planks forming a marker, which are secured with metal stakes, the stakes holding the planks firmly to the ground. Once you see the square marker, you must walk *only* diagonally from the south-east corner towards the north-west corner as indicated in the diagram that I

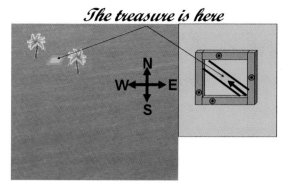

Fig. 4.30 The map and the square lot according to the king's description.

Fig. 4.31 This is what I found.

sent you (Fig. 4.30). When you reach the center of the square — you should dig exactly at the center — you will find your treasure, and it is yours for the taking. To ensure that you and only you will reach the treasure, I have put landmines in the entire area between the planks. You should be careful not to step outside the strip of land that is about 10 cm wide, and go all the way diagonally from one corner to the second corner of the square. If you are not careful, a landmine will explode and you will be blown to smithereens. But if you follow my instructions as explained and shown in the diagram, you are going to be supremely rewarded.

What a hoax, I thought to myself, but then again, what if it was for real? Wanting to satisfy my curiosity as well as my love

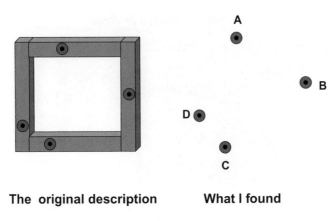

The original description   **What I found**

Fig. 4.32 This is what was left.

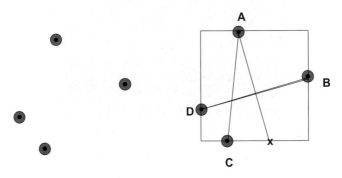

Fig. 4.33 The solution is based on a geometric theorem. Two perpendicular lines connecting opposing edges of a square have equal lengths. This is easy to prove. Once you have the two lines AC and BD, you draw a line from the point A, perpendicular to BD. You get the point x. Then construct the square by drawing the lines as indicated in the next figure.

for adventure, and with a first-class ticket to boot courtesy of the King, I decided to go for it.

I flew to Goldland, and from the mainland, I took a boat that ferried me to Fortune Island. Before we left the wharf, I told the boatman exactly where I wanted to go. With much

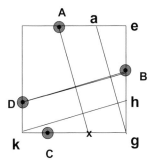

Fig. 4.34 Hint for the proof of theorem. Draw two lines: ag and kh, parallel to Ax and DB respectively, as shown in the figure. Show that the two triangles kgh and gea are congruous.

effort, he spoke to me in halting English, "What for to go to island? Nothing, nothing in place, no beauty, no thing, no one." I said it was okay and I assured him that it was worth seeing. Scratching his head, we headed off to the blue waters until we finally reached the northern part of the island.

I giddily walked towards the spot where the two palm trees stood. The King had so far spoken the truth. There were indeed two majestic looking palm trees that gracefully swayed with the wind. Much to my chagrin, the wooden marker was no more, being burned to cinders. However, the four stakes were still there. All I knew was that each of the stakes was fastened to one of the sides of the square, but I could not figure out where the square lot was and how to find the diagonal path that would lead me safely to the center of the square.

Can you help me in finding the treasure? If you do, I promise to share the treasure with you. Since you are now acquainted with probability, perhaps you can tell me what the probability is of getting the treasure if I just walked into the area in between the four stakes and dug at a random point in that area?[21]

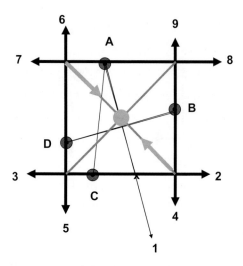

Fig. 4.35 Reconstruction of the square. Draw the lines 1, 2, 3, . . . , to 9, in this order, to get the original square.

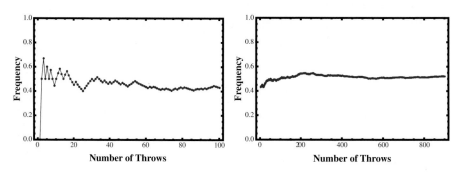

Fig. 4.36 Frequency of outcomes "H" in throwing a coin, for small and large number of throws.

Before we conclude this session, I invite you to look at Fig. 4.36. I plotted here the fraction of Hs as a function of the number of throws. Note that after about 200 throws, the fraction is almost constant at about 1/2.

4.10 Conclusion

This session was quite long. However, it contains the most important concepts in probability theory. We have learned what conditional probabilities are and how to calculate them. In the next session, we shall discuss some real-life examples. You will realize that in order to understand these examples, and to avoid any hazardous pitfalls, you have to understand what conditional probabilities are.

Session 5

Bayes' Theorem and Its Applications

Bayes' theorem is very simple and is very useful for solving some probability problems, the answers to which are sometimes counter-intuitive. Do not recoil upon hearing the word "theorem." If you have absorbed the concept of conditional probability, you should have no difficulty in understanding Bayes' theorem. It is nothing more than a small variation on the theme of conditional probability. Yet this theorem is extremely useful in many applications. Unfortunately, it is sometimes misused and sometimes abused. Understanding this theorem is very satisfying and rewarding. To achieve a full understanding, you should read this session carefully, check my arguments and do the exercises.

As we have done in the previous sessions, we shall start by a simple example that will lead you to derive a special case of the theorem. Then we shall generalize the theorem, examine a few useful applications and entertain ourselves with one example of an abuse of the theorem.

Before doing so, let me tell you a little story in which a probabilistic mis-judgment could have disastrous consequences.

A man lived in a very clean, hygienic and environmentally healthy city. It was *known* that no one in the city was a carrier of HIV. One day, he took a very reliable test for HIV. He took the test out of mere curiosity, as he was sure that he was healthy and the tests would yield negative results. Upon hearing the news that the results were positive, he was terrified by the fact that he would soon be sick, and would probably die after a protracted battle with the disease, suffering an agonizingly long period of excruciating pain. Therefore, he decided to commit suicide.

Can you help him postpone his decision and assess his medical situation more carefully? Perhaps the bad news was not as bad as he thought. I am sure that he would not have reached this decision had he known about Bayes' theorem! I urge you to read this session carefully. It might just save your life one day!

Before we undertake the study of Bayes' theorem, please read the following two *real* stories.

5.1 Newsworthy News Crazy

I read in one of the local newspapers some very disturbing statistics: Apparently, 25% of Israeli women suffer from depression, while 17% of Israeli men suffer from the same fate. The reporter concluded that 42% of Israel's population suffers from depression. Does this alarming news mean that everyone in Israel should go through a check-up?

5.1.1 *50% off or a rip-off?*

A few years back, I got in touch with a reputable real estate agency and told them of my intentions to purchase a new apartment. A young and seemingly inexperienced sales agent took me to several nice apartments, one of which I really liked.

I had almost decided to buy the apartment, when the agent handed me a contract that stated that the real estate company was entitled to a 15% commission from the apartment's total value.

I thought that 15% was rather too high and so I tried to haggle with the agent. Since I was going to sell my old apartment anyway, I thought it might be a good bargaining chip. I told the lady that if she agreed to lower their commission, I would sell my old apartment and buy a new apartment through their agency.

Probably feeling confident that a deal was in the works, the agent said, "Why, yes, of course. We would be glad to sell your apartment. If you sell your apartment through us and buy a new apartment through us, you are entitled to 50% discount on the commission."

I thought that this was rather generous and so I readily acceded to her proposal. In an instant, a calculator materialized out of nowhere, and before she punched the keys, she apologized for her weakness in mathematics. "Oh, mathematics is my Waterloo. I was never good with numbers," she declared. I nodded. A few key punches here and there and then she said, "Okay, 15% commission on the buying, and 15% commission on the selling, altogether 30%, right? So with a 50% discount on the commission, you will only pay 15% if you buy and sell with us."

What a generous offer indeed, but I had to decline her offer.

On the way home, I had wondered if the agent was perhaps the same person who wrote the article in the newspaper on the statistics on Israelis suffering from depression. Your guess is as good as mine.

Let us warm up with a simple example. I throw a dart at board A. The board is divided into two regions: A_1 and A_2 (Fig. 5.1), such that the union of the two regions is equal to the entire region of the board. We write this as $A = A_1 \cup A_2$.

I tell you that the dart hit the board, and that all of the points on the board are equivalent. As always in this book, we shall not be interested in the exact mathematical *point* at which the dart

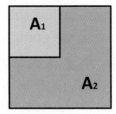

Fig. 5.1 A board divided into two regions, A_1 and A_2.

hit the board. There is an infinite number of points. Instead, we shall assume that the board is divided into very small squares, as small as you wish, but the total number of these is finite. In addition, we neglect the possibility that the dart hit the dividing line between the two regions A_1 and A_2 (Fig. 5.1).

What is the probability that the dart hit the region A_1?

We assume that the probability of hitting any region on the board is simply the fraction of that area from the total area of the board. If the total area of the board is a, and the areas of A_1 and A_2 are $a_1 = \frac{a}{4}$ and $a_2 = \frac{3a}{4}$, respectively, we assume that the probability of the "event" A_1 (i.e. hitting the region A_1) is:

$$Pr(A_1) = \frac{a_1}{a} = \frac{1}{4}$$

and likewise the probability of the "event" A_2 is:

$$Pr(A_2) = \frac{a_2}{a} = \frac{3}{4}.$$

Think about these probabilities as an example of the application of the *classical definition* of probability. Here, we have an infinite number of "elementary" events, but we can view the total board in Fig. 5.1 as divided into four regions of equal areas. We are interested in which one of the small squares the dart hit. Now we have a finite number of elementary events, and the two probabilities written above are obtained by the classical definition of probability (see Session 2).

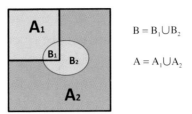

Fig. 5.2 The same board as in Fig. 5.1 but with a region B painted blue.

Now that we know the probabilities of the two events A_1 and A_2, let us introduce a modification of the problem. Suppose that I have painted with blue some region on the board, denoted B in Fig. 5.2. You are told that 10% of region A_1 was painted and 20% of region A_2 was painted.

What is the percentage of the total area of the board that was painted?

I hope that the two stories I have told you will dissuade you from rushing to the conclusion that if 10% of A_1 was painted and 20% of A_2 was painted, then the percentage of the total area that was painted is $10\% + 20\% = 30\%$! Let us first calculate the total area on the board that is painted.

The area of A_1 that is painted is a_1. Therefore, the painted area of A_1 is $\frac{10}{100}a_1$, and the painted area of A_2 is $\frac{20}{100}a_2$. The *total* painted area is therefore:

$$b = \frac{10}{100}a_1 + \frac{20}{100}a_2.$$

The fraction of the total board painted in blue is also the probability of hitting the region B. Therefore:

$$Pr(B) = \frac{b}{a} = \frac{\frac{10}{100}a_1 + \frac{20}{100}a_2}{a} = \frac{10}{100}Pr(A_1) + \frac{20}{100}Pr(A_2)$$

$$= \frac{1}{10} \times \frac{1}{4} + \frac{2}{10} \times \frac{3}{4} = \frac{7}{40}$$

i.e. the total painted area is 17.5% of the board, which is quite different from the quick estimate we obtained by summing 10% and 20%.

Next, we ask what the conditional probability is of hitting the painted area B *given* that the dart hit A_1. Given that A_1 occurred, the probability of hitting the painted area is simply the fraction of A_1 that is painted:

$$Pr(B|A_1) = \frac{1}{10}.$$

Again, you can derive this result from the classical definition of probability. The total area of A_1 is a_1, and the painted area in A_1 is $\frac{10}{100}a_1$; therefore, the fraction of the painted area in A_1 is $\frac{10}{100} = \frac{1}{10}$.

Likewise, we can calculate the probability of hitting the painted area in A_2 as:

$$Pr(B|A_2) = \frac{2}{10}.$$

At this point, we have all of the required probabilities:

$$Pr(A_1) = \frac{a_1}{a}, \quad Pr(A_2) = \frac{a_2}{a} \tag{5.1}$$

$$Pr(B|A_1) = \frac{1}{10}, \quad Pr(B|A_2) = \frac{2}{10} \tag{5.2}$$

These are sometimes referred to as the *a priori* probabilities, i.e. these are given in advance, and later we shall calculate *a posteriori* probabilities.

Bayes' theorem is concerned with calculating a kind of "inverse" probability. You are given the probabilities as in (5.1) and (5.2), and you are asked about the *inverse conditional* probabilities. For example, given that the dart hit the painted area B, what is the probability that it is in area A_1? In other words, we want to find the conditional probabilities $Pr(A_1|B)$ and $Pr(A_2|B)$. You can see that in the latter probabilities, the roles of A_1 and B (and of A_2 and B) are "reversed" compared with their roles in (5.2). In other words, we reverse the roles of the *condition* and the outcome events.

To calculate $Pr(A_1|B)$, we use the definition of the conditional probability (see Session 4):

$$Pr(A_1|B) = \frac{Pr(A_1 \cap B)}{Pr(B)}.$$

Remember that $Pr(A_1|B)$ is the probability of A_1 given B, and $Pr(A_1 \cap B)$ is the probability that A_1 and B occurred.

Next, we want to rewrite the right-hand side of the equation in terms of the quantities we already know. Using again the definition of conditional probability, we have:

$$Pr(A_1 \cap B) = Pr(A_1)Pr(B|A_1) = \frac{a_1}{a} \times \frac{1}{10} = \frac{1}{40}.$$

Note carefully that the quantity $Pr(B|A_1)$ is the probability of B *given* A_1. Furthermore, we already calculated the probability of hitting B, which we can rewrite as:

$$Pr(B) = Pr(B \cap A_1) + Pr(B \cap A_2)$$
$$= Pr(B|A_1)Pr(A_1) + Pr(B|A_2)Pr(A_2)$$
$$= \frac{1}{10}\frac{a_1}{a} + \frac{2}{10}\frac{a_2}{a} = \frac{1}{10} \times \frac{1}{4} + \frac{2}{10} \times \frac{3}{4} = \frac{7}{40}.$$

Please follow the steps carefully. The first line of the equation means that the region B is simply the sum of the areas that B "cuts out of A_1" and the area that B "cuts out of A_2" (see the two areas B_1 and B_2 in Fig. 5.2). This equality is sometimes referred to as the *total probability theorem*. As you can see, this theorem is nothing but a statement that the area B in Fig. 5.2 is the sum of the areas B_1 and B_2. Next, we wrote each of the joint probabilities in terms of the conditional probabilities, and in the final step, we plugged in the numbers we already had. The result we need is therefore:

$$Pr(A_1|B) = \frac{\frac{1}{40}}{\frac{7}{40}} = \frac{1}{7}.$$

Check this result by looking at the various areas in Fig. 5.2.

If you have followed me so far in calculating the required probability, you have just *derived* Bayes' theorem. There is nothing more in Bayes' theorem than what you have already seen. It is simply rewriting the conditional probability in a slightly more complicated, albeit more useful way. However, if you still have doubts, let us do another simple example that is also important in the quality control of products in different factories. We shall repeat exactly the same calculation but using a different "language."

There are two factories A_1 and A_2 producing and supplying the same product, say computers, to the same store. The store received 40 computers, 10 from factory A_1 and 30 from factory A_2. If I pick a computer at random in the store, and I know that it was produced in either A_1 or A_2, what is the probability that it was produced in A_1? What is the probability that it was produced in A_2? These probabilities are:

$$Pr(A_1) = \frac{10}{40} = \frac{1}{4}$$

$$Pr(A_2) = \frac{30}{40} = \frac{3}{4}.$$

It is also known from collected data over many years that, on average, about 10% of the computers produced by A_1 are defective, and about 20% of the computers produced by A_1 are also defective.

Write down the probability of finding a defective computer (event D) if you *know* that it was produced by A_1. Write down the probability of finding a defective computer if you *know* that it was produced by A_2. Denote these as $Pr(D|A_1)$ and $Pr(D|A_2)$, respectively.

Next, write down the probability of picking up a defective computer in the store when you *do not know* from where it was produced.

Finally, calculate the following "inversed" problem. You go to a store, pick up a computer at random, take it home and find that it is defective. What is the probability that this computer was produced in A_1, and what is the probability that it was produced in A_2?

I will let you do this exercise by yourself.[1] Then we shall formulate the theorem for the more general case.

Now that you have solved the problem for this particular example, let us write the theorem in general terms.

You are given n events, $A_1, A_2, \ldots, A_n$. All are *disjoint* events, which means that for each pair of events A_i and A_j, their *intersection* is empty.

$$A_i \cap A_j = \emptyset, \quad (i \neq j).$$

For an example, see Fig. 5.3. We also denote by A the *union* of all the events $A = \cup_{i=1}^n A_i$. Let B be any event (not the empty event, i.e. B $\neq \emptyset$) that intersects with all or part of the events $A_1, \ldots, A_n$. You are given the *a priori* probabilities:

$$Pr(A_i) = p_i \qquad (5.3)$$
$$Pr(B|A_i) = q_i \qquad (5.4)$$

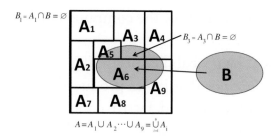

Fig. 5.3 A board divided into nine regions of different areas. A region B is painted in grey.

where p_i is the probability that the event A_i occurred and q_i is the conditional probability that B will occur if it is known that A_i has occurred. In Fig. 5.3, the entire region B is shown in grey. In this particular example, we see that B does not intersect with A_1, while B_3 is the intersection of B with A_3, i.e., some of the A_i regions have non-empty intersection with B, and some do not.

Now you are asked to write down the probability that A_i occurred when you know that B has occurred. The answer is straightforward. We start from the definition of the conditional probability:

$$Pr(A_i|B) = \frac{Pr(A_i \cap B)}{Pr(B)}. \tag{5.5}$$

We rewrite the numerator using the definition of conditional probability, but in reversing the roles of A_i and B:

$$Pr(A_i \cap B) = Pr(B|A_i)Pr(A_i). \tag{5.6}$$

Can you guess why we did this?

Then, we rewrite the denominator of (5.5) in a more complicated way:

$$Pr(B) = \sum_{i=1}^{n} Pr(B \cap A_i) = \sum_{i=1}^{n} Pr(B|A_i)Pr(A_i). \tag{5.7}$$

This simply means that if B intersects with some or all of the events A_i, then the probability of B is the sum of the probabilities of all of the intersections $B \cap A_i$. This is the *theorem of total probability*.

Remember that we assumed that all of the A_i regions are disjoint events. This means that the probability of the union of all A_i (denoted A) is the *sum* of the probabilities of each A_i. Convince yourself that if all A_i are disjoint events, then all the intersection events $B \cap A_i$ are also disjoint events (see Fig. 5.3).

Now, rewriting (5.5) with the help of (5.6) and (5.7), we get the famous Bayes' theorem:

$$Pr(A_i|B) = \frac{Pr(B|A_i)Pr(A_i)}{\sum_{i=1}^{n} Pr(B|A_i)Pr(A_i)}. \qquad (5.8)$$

This might look complicated, but as (I hope) you have convinced yourself by working through the previous examples, this theorem is nothing but a rewriting a conditional probability in a more complicated, albeit much more useful form.

Why is this theorem so useful? You will see why from the following examples. Here, you have to look carefully at the last formula and see that on the right-hand side of the equation there are all of the quantities we *know*. On the left-hand side is a quantity we want to calculate. The known quantities on the right-hand side are sometimes referred to as *a priori* probabilities (given beforehand). The quantity on the left-hand side is referred to as the *a posteriori* probability (given after). The reason is that in the *a priori* case we go from A_i to B. In the *a posteriori* case, we go in the "reverse" direction from B to A_i.

Be careful not to confuse the "direction" from A_i to B or from B to A_i as directions *in time of occurrence*, nor should you confuse conditional probability with cause and effect. If you have any doubts about these "directions" in time, go back to Session 4, especially the examples at the end of the session.

We now turn to some important applications of Bayes' theorem, but before doing these "serious" examples, let us relax and examine the following "promising" business offer.

5.1.2 *A banana business proposition*

A few years ago, I was presented with a sound business proposal by a good-looking, well-dressed gentleman who introduced himself as the proprietor of banana plantations in Mexico. His proposal was for me to buy from him 1000 bananas for $1 apiece. He projected that in the city where I lived, which contained about a million people, I would easily earn a lot of money if I followed his instructions as follows: Select at random 1000 people from the entire population of the city, and give one banana to each of them. The deal is simple. If a person received a banana and liked it, he or she will pay me $4. If the person did not like the banana, he or she will pay me only $1.

The offer seemed reasonable. I trusted that the businessman knew that people are honest by nature, and that they would certainly pay me $4 if they liked the bananas. I accepted the offer initially and bought 1000 bananas. I randomly distributed them just as he had suggested to 1000 people. After a few days, about 600 persons came back to me to tell me they did not like the bananas and paid me $1 each, or a total of $600. The 400 who liked the bananas paid me $4 each, or a total of $1600. Thus, from a $1000 investment, I had earned $1200. Not bad at all for a start! Buoyed by my substantial earnings, I ventured yet again the following month, and bought another 1000 bananas for the price of $1000. Just like the first time, I distributed the bananas randomly to 1000 persons, and waited for a few days.

Four hundred people came back and said they did not like the bananas and I was paid a total of $400 from them. Six hundred of them, however, liked the bananas and they paid me $4 each, or a total of $2400. My second investment proved to be even better, gaining me bigger earnings. From my $1000 investment, I received $2800, or a net profit of $1800.

A few months later, the same man came up with an extraordinary business proposal. He told me that he had discovered a beautiful island inhabited by 1,000,000 very rich people. I was to apply the same scheme; select 1000 people, give each one of these 1000 persons a banana, and after a few days they would come back and pay me, depending on whether or not they liked the bananas. However, there was a twist to his proposal. Those who did not like the bananas were to pay me (as usual) $1, but those who liked them were to pay me a staggering $1000 apiece. He assured me that the inhabitants of the island were honest, and that they would most certainly pay me if they liked the bananas. Another twist to his proposal was that instead of me buying the bananas for only $1 apiece, I was to pay him $4 apiece. In addition, I would have to pay the cost of transporting the bananas, which would cost roughly $5000, plus my plane ticket costing $1000 more.

Doing a quick mental calculation, I figured that altogether I was going to invest $4000 for the bananas alone, $5000 for transporting them, and an additional $1000 for my flight ticket, or a total of a $10,000 investment. Based on my previous experience, I calculated that about half of the 1000 people would like the bananas, while the other half would not, in which case I would get $500 from those who did not like the bananas, and 500 × $1000 from those who liked them. This would be a huge profit. But what concerned me was whether only a quarter or maybe 1% of the population liked the bananas, which would mean that I was bound to lose money on all of those who did not like them. If only 1% liked the bananas, I would recoup $990 from those who did not like them, and 10 × $1000 from those who did like them. Based on my calculation, I would still make a profit, although only by a small margin.

While I hesitated and pondered on his proposal, he came up with yet another offer. He said that he knew that I was not really keen on the idea of distributing the bananas *randomly* because I would lose $3 apiece if people did not like them, and so he offered me a device that he had developed — a simple sensor that looked like a thermometer. The sensor worked in such a way that when it was pointed at someone who liked bananas, a green light would turn on. He told me that he had tested the device on a large population and found that the green light turned on in as many as 99% of his subjects who liked bananas. He also found out that the sensor showed a false green light for only 0.001% of those who did not like bananas. The detector seemed quite reliable and was worth buying, even though it would cost me an additional $1000. After convincing me of the reliability of his device, he told me a different aspect of the scheme. Whereas before I had distributed the bananas randomly to 1000 persons, now I could use the device to *test* the 1000 people I was to distribute the bananas to. Whenever the green light turned on, that would be my signal to give a person a banana. This scheme assured me that I would not lose money by giving bananas to too many people who might not like them. The reliability of the detector meant that only very few of the people to whom I would give a banana would not like it.

I was blown away by the sensor and I appreciated the man's concern about me losing money if I were to make the mistake of giving bananas to people who did not like them. To cast away any doubts I might still have had, he put all of the conditions in writing in a contract. He repeatedly assured me that all of the inhabitants of the island were honest, and that if a person liked their banana, he or she would pay me $1000. To further entice me, and to erase all my doubts, he added a provision in the contract stating that if people liked the bananas, but some stroke of bad luck, such as illness, death or force majeure, rendered them incapable of paying me, he, Mr. Banana Seller, would instead pay the $1000 that was due to me.

All of this seemed to be promising, and I did not see any possibility that I would lose money. It seemed like a foolproof plan. I thought about all of the possible outcomes and I could not find any reason why I should decline his proposal. I promised to think about the proposal overnight, and to sign the contract on the next day.

When I got home and I sat down to read, I got a hold of Arieh Ben-Naim's book (2008) in which Bayes' theorem was discussed. The next day, I declined the offer.

Can you guess why I changed my mind?

Would you accept such an offer? If so, why would you accept the offer? Or if not, why not?

I will give you my answer after you read the next section. Perhaps you might not even need an answer. Whatever your answer is, read the following section very carefully. It is one of the most important applications of Bayes' theorem. Some of the results can be very surprising.

5.2 False-Positive Results

A very important application of Bayes' theorem is in assessing the reliability of a test for a certain disease. Suppose that a given country, say Virusland, has a substantial fraction of its population carrying a virus that causes a serious disease. Let us assume that a fraction r of the total population are carriers (C) of the virus, and the other fraction of the population $(1 - r)$ are not. This means that when you randomly meet a person in Virusland, the probability that he or she is a carrier is:

$$Pr(C) = r$$

and the probability that he or she is not a carrier (NC) is:

$$Pr(NC) = 1 - r$$

Figure 5.4 shows schematically the two regions C and NC.

Suppose that we have a test (T) that is very sensitive to detecting the virus. It is not perfect, but it is highly reliable. The collected statistics over several years show that when the test is applied to a *carrier*, 99% of the tests are positive. We write this as:

$$Pr(T^+|C) = p = 0.99.$$

This equation reads: The conditional probability that the test is *positive* (T^+), *given* that the person is a carrier (C), is very high (0.99).

However, as we have said, the test is not perfect. Sometimes it fails to show the correct result. It was found that when the same test was applied to persons who are *known* to be non-carriers,

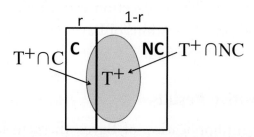

Fig. 5.4   Schematic regions of carriers (C) and non-carriers (NC).

the test also shows positive results, but only in a very small percentage of the cases. Let us say about one person in 100 non-carriers showed positive results. We write this as:

$$Pr(T^+|NC) = q = 0.01.$$

The equation reads: The conditional probability that a test is *positive* (T^+), *given* that the person is a non-carrier (*NC*), is very small (0.01).

The experimental data are impressive indeed. The probability of correct results is very high and the probability of incorrect results is very small. Note that I chose $q = 0.01$, but I could have chosen any other small number, like $q = 0.001$ or $q = 0.0001$. This number does not have to be related to the probability $Pr(T^+|C)$. I chose the values of 0.99 and 0.01 for convenience in the following calculations.

Now, we want to ask the *inverse* question. Suppose you live in Virusland and you do not know whether you are a carrier or not. You decide to take the test, and after two days, the lab manager tells you that the result was *positive*. That is obviously unpleasant news. But is it very bad news, or extremely bad news? In other words, *given* that the result is positive, what is the probability that you are a carrier? You might be surprised to learn that sometimes there is no reason to be alarmed by the results. In order to convince yourself read the following very carefully.

Let us use Bayes' theorem in order to calculate the probability of being a carrier, or a non-carrier, *given* that the test is *positive*. We write these two probabilities as follows:

$$Pr(NC|T^+) = \frac{Pr(T^+|NC)Pr(NC)}{Pr(T^+)} \quad (5.9)$$

This is the probability of being a non-carrier (*NC*) *given* a positive test (T^+).

Similarly, the probability of being a carrier given a positive test is:

$$Pr(C|T^+) = \frac{Pr(T^+|C)Pr(C)}{Pr(T^+)}. \qquad (5.10)$$

$Pr(T^+)$ is the probability of getting a positive result when the test is applied to the entire population (see the total grey area in Fig. 5.4).

We write the total probability theorems as follows:

$$Pr(T^+) = Pr(T^+ \cap C) + Pr(T^+ \cap NC)$$
$$= Pr(T^+|C)Pr(C) + Pr(T^+|NC)Pr(NC).$$

In the first equality on the right-hand side, we wrote the probability of "test positive" (T^+) as the sum of the probability of "test positive *and* carrier" and the probability of "test positive *and* non-carrier." This is simply the *total* grey area in Fig. 5.4, written as a sum of terms as shown in the figure. Now we have all of the data on the right-hand side of this equation. We plug the numbers we have into Equations (5.9) and (5.10) and get:

$$Pr(NC|T^+) = \frac{0.01 \times (1 - r)}{0.99 \times r + 0.01 \times (1 - r)} \qquad (5.11)$$

$$Pr(C|T^+) = \frac{0.99 \times r}{0.99 \times r + 0.01 \times (1 - r)}. \qquad (5.12)$$

I have deliberately left the fraction r of carriers in the population unspecified in the last expression. I want to surprise you again with a very unexpected and counter-intuitive result. But before doing this, suppose for concreteness that, in

Virusland, the fraction of the carriers in the entire population is about 0.5.

In this case, the two conditional probabilities in Equations (5.11) and (5.12) are (with $r = 0.5$):

$$Pr(NC|T^+) = \frac{0.01 \times 0.5}{0.99 \times 0.5 + 0.01 \times 0.5} = 0.01$$

$$Pr(C|T^+) = \frac{0.99 \times 0.5}{0.99 \times 0.5 + 0.01 \times 0.5} = 0.99.$$

The last result means that if you tested positive (T^+), then the chance of you being a carrier (C) is 99%. This is clearly consistent with what you might have expected from such a reliable test. If your test is positive, this is very sad news indeed, and you should be alarmed.

But suppose you come from a country where it is *known* that only 1% of the population are carriers of the virus. You must be fortunate to live in that country. You take the same test with the same data as given above, and you find to your surprise that the test is positive. Now the question is what is the probability of obtaining a false-positive result, knowing that only 1% of the population are carriers? This is the reason that I left the fraction r unspecified in the formula above. Plug $r = 0.01$ into Equations (5.11) and (5.12) and you will determine your chances of being a carrier given that you tested positive. The two conditional probabilities are now:

$$Pr(NC|T^+) = \frac{0.01 \times 0.99}{0.99 \times 0.01 + 0.01 \times 0.99} = \frac{1}{2}$$

$$Pr(C|T^+) = \frac{0.99 \times 0.01}{0.99 \times 0.01 + 0.01 \times 0.99} = \frac{1}{2}.$$

This means that if the test is positive, you have only a 50% chance of being a carrier. This is bad news, but far less alarming news than in the previous case. But how come I obtained a very different result using *exactly the same test* that is highly reliable?

The reason behind this surprising result is the same reason that I left the fraction r in the formulas (5.11) and (5.12) unspecified. These formulas show that both $Pr(NC|T^+)$ and $Pr(C|T^+)$ *depend on the fraction* of carriers in the *entire population* of the country that you live in.

Here is a shocking and an even more surprising result for you to ponder on.

Suppose you live in a country with a high standard of living and good medical care, and only 0.1% of the entire population is *known* to be carriers of the virus in question. Of course, you are lucky to live in such a country. If you pick a person at random in this country, the chance that this person is a carrier is only 0.001, which is very small indeed.

One day, you decide to take the test for the virus (the same test with the same reliability as given above). You are confident that the result will be negative since it is so rare in your country. But lo and behold, two days after you took the test, you are told the *shocking news* that the result is *positive*. Your mood plummets, and you rush to the doctor to seek help.

The doctor checks you over, looks at the results and tells you that testing positive for such a nasty disease is indeed bad news, but that the chances of you actually being a carrier are quite low.

How could the doctor dare to say this, having just seen the bad news that the test was positive?

The answer is that the doctor, unlike the patient, knew about Bayes' theorem. He quickly calculated the chances of a false-positive result and found that:

$$Pr(NC|T^+) = \frac{0.01 \times 0.999}{0.99 \times 0.001 + 0.01 \times 0.999} = \frac{999}{1098}$$

$$Pr(C|T^+) = \frac{0.99 \times 0.001}{0.99 \times 0.001 + 0.01 \times 0.999} = \frac{99}{1098}.$$

Now you see that the chances of a false-positive result are about $\frac{999}{1098} \approx 0.9$, and the chances that you are a carrier, even though you tested positive, are about $\frac{99}{1098} \approx 0.1$. You see that, in this case, there is some reason for concern, but not too much.

Remember the guy from the beginning of this session who was about to commit suicide? Well, can you now tell him that he has *no reason* to be alarmed by his positive test result? If you are not sure, read the following.

Figure 5.5 shows how the probabilities $Pr(C|T^+)$ and $Pr(NC|T^+)$ change as a function of the fraction of carriers in the population. You can see that both of these *conditional probabilities* can have values ranging between zero and one.

In particular, the probability of a false-positive result can also be nearly one. This conclusion is counter-intuitive. It means

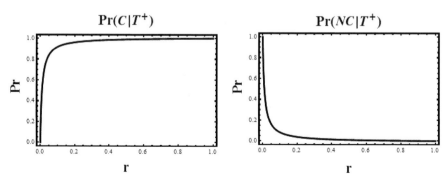

Fig. 5.5 Plot of the function $Pr(C|T^+)$ and $Pr(NC|T^+)$ as a function of r.

that no matter how reliable the test is, say $Pr(T^+|C) = 99\%$ or 99.99%, there is always a possibility of a very large probability of obtaining a false-positive result. As you can see from Fig. 5.5, this will occur when the fraction of carriers (r) is nearly zero. The reason it is counter-intuitive is that we tend to confuse $Pr(T^+|C)$ with $Pr(C|T^+)$. The first probability means the probability of getting a positive result (T^+) when applying the test to a person who is *known* to be a *carrier*. This quantity depends on the statistics one has accumulated in using this test. It can be extremely reliable and it can be nearly one (this is because the test was applied to people who were *known* to be carriers by some other method). Therefore, as the test is so reliable, one tends to conclude that if you tested positive, then it is almost certain that you are a carrier. Here is where our intuition fails us.

The reason that our intuition fails us is that while $Pr(T^+|C)$ is *given*, the quantity $Pr(C|T^+)$ *depends on the fraction of the carriers in the population.* As the plot in Fig. 5.5 shows, this quantity can change from zero to one. This means that in an extreme case, you might test positive, but you can be almost sure that you are not a carrier. Look again at the two curves in Fig. 5.5. Tell yourself in *words* what each of these curves means. Try also to understand qualitatively why these curves have these forms.

To understand this result better, think of a city or country that is *known* to have *no single carrier* in its entire population. If you apply the test to any person in this population, you will always find:

$$Pr(C|T^+) = 0, \quad Pr(NC|T^+) = 1.$$

Thus, no matter how reliable the test is, if you test positive, you should be certain that the result is false, i.e. the probability that you are a carrier is zero! Note that I used the word "certain" because it is *given* that $r = 0$, i.e. all of the people in this city or country are not carriers.

At the other extreme, suppose you have a city or country in which it is *known* that every single person is infected with the virus. You take a test, and no matter how lousy the test is, and no matter how reliable the statistics are about the test, if you get a *negative* result, you can be sure it is a false-negative. This means that your test is negative, but the probability that you are a carrier is one![2]

Exercise: Consider the following problem: A metal detector in an airport goes off in 999 out of 1000 of the cases in which a person carrying metal objects passes through, i.e. $Pr(T^+|C) = \frac{999}{1000}$. There are also false-alarm cases in which the detector goes off even if the passenger is *not carrying* any metallic objects, i.e. $Pr(T^+|NC) = 10^{-6}$. Given that the detector goes off, what is the probability that the passenger is a carrier of a metallic object?

To answer this question, you have to repeat the same calculations as we did above. It is a good opportunity for you to check if you understood the "false-positive" section.

Now I owe you an answer as to why I changed my mind about the banana business.

I forgot to ask Mr. Banana Seller what fraction of the population of the island likes bananas. As in the example above, one would not suspect the dependence of the "false-positive" on the fraction of the carriers in the population. In the present case, a "carrier" is replaced by a "banana-liker." Normally, one would assume that about 50% of the population like bananas, and 50% do not. But what if the *entire* population does not like bananas? In this event, even if everything that was promised by the banana dealer was true, and the detector was perfect, nothing would help if *no-one* on the island likes bananas. You will end up losing money.

5.3 The Three Cards Problem: Use and Misuse of Bayes' Theorem

Here is a seemingly easy problem in which it is very easy to misuse Bayes' theorem. Please go through the arguments very carefully. This will prepare you for a far more serious problem — a decision of life and death.

You are shown three cards. One is colored blue on its two sides, call it BB, one is colored green also on both sides, call it GG, while the third is colored blue on one side and green on the other. Let us call it BG (Fig. 5.6). You close your eyes and pick one card at random from the box, and put it on the table. It is assumed that the probability of picking any one of the cards from the box is the same, i.e. 1/3.

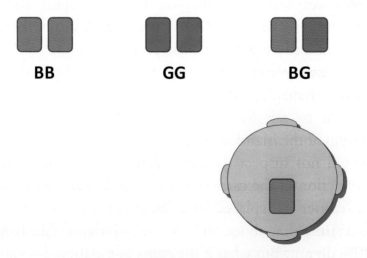

Fig. 5.6 The three cards problem.

We write this *given* information as:

$$Pr(BB) = Pr(GG) = Pr(BG) = \frac{1}{3}.$$

The total number of blue-colored sides in all three cards is three, and the total number of green-colored sides is also three. Therefore, when you open your eyes the probability of seeing the color blue on the card on the table is 1/2, and the probability of seeing the color green on the table is also 1/2. See Fig. 5.6 and convince yourself of these probabilities.

Now, when you open your eyes, you see that the card you chose and laid on the table has the blue color facing upwards. You cannot see the color of the other side of that card. What is the probability that the other (unseen) side is also blue? In other words, we need to calculate the *conditional probability* that the other (unseen) side of the card on the table is blue, *given* that you see that the upper side is blue. This sounds like a very simple problem. But wait and see — perhaps there is a trap lurking.

A quick answer is the following.

Given the fact that there are three cards each having an equal probability of being picked (1/3), the probability that the other side will also be blue means that you have picked a BB card. The probability of picking the BB card is as written above, simply 1/3.

Do you agree with these results?

Let me try to give you another quick answer. Given that you *see* a blue color on the face of the card means that the card you have picked is *not* GG. There remain two possibilities for the card you have picked: Either BB or BG. Since these two events have equal probabilities, the required probability of BB is 1/2.

Are you happy with this answer?

Write down your preferred solutions based on what I offered above, or perhaps you have another answer.

Let us solve the problem mathematically.

5.3.1 *(I) First solution*

Given that you see the blue color of the upper side of the card on the table (call it B_1), this certainly excludes the possibility that this card is GG. Therefore, it must be either BB or BG. We write this event $BB \cup BG$ (reads: either BB *or* BG). We now use the definition of the conditional probability of the event BB, given that we *know* that the result is $BB \cup BG$:

$$Pr(BB|BB \cup BG) = \frac{Pr(BB \cap (BB \cup BG))}{Pr(BB \cup BG)}. \qquad (5.13)$$

This looks awkward but it is simple to read. $Pr(BB \cup BG)$ is the probability of getting either BB or BG. The probability of this event is 2/3 (two cards out of three). The event $BB \cap (BB \cup BG)$ is exactly the same as the event BB. It is easy to see that the intersection of any event A with an event that includes A is the same as A itself (Fig. 5.7). In other words, the occurrence of either BB *and* (GG or BB) is the same as the occurrence of BB. We also know that the probability of BB is 1/3. If you are not sure of this, look at Fig. 5.8. Hitting either region L or region R

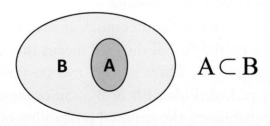

Fig. 5.7 An event A is included in the event B.

$$L \subset L \cup R$$
$$L = L \cap (L \cup R)$$

Fig. 5.8 Check the two relationships under the figure.

is $L \cup R$, i.e. the total region. Hitting region L *and* $L \cup R$ is the same as hitting the left region of the board.

In our problem, we know the probabilities on the right-hand side of Equation (5.13). Therefore, plugging in these probabilities, we get:

$$Pr(BB|BB \cup BG) = \frac{Pr(BB)}{Pr(BB \cup BG)} = \frac{\frac{1}{3}}{\frac{2}{3}} = \frac{1}{2}. \qquad (5.14)$$

This seems to prove the second of the quick answers that I gave you above. But hold on, do not rush to conclude that you have now got the right answer. Check the next solution; perhaps you might change your mind. I suggest that you pause and think whether all of the arguments given above are correct. There is one suspicious step — can you find it?

5.3.2 *(II) Second solution*

Again, in this second solution, we use the conditional probability in order to formulate the problem, but in slightly different terms. We are asked for the probability of the second side of the card, call it B_2, being blue given that the first side, call it B_1, is blue.

We write this as:

$$Pr(B_2|B_1) = \frac{Pr(B_2 \cap B_1)}{Pr(B_1)} = \frac{Pr(BB)}{Pr(B_1)}. \qquad (5.15)$$

Clearly, the event B_2 *and* B_1 is the same as BB. Look at Equations (5.13) and (5.15). Interpret them in *words*. Are these two formulations of the problem identical?

Now, we already know $Pr(B_1) = 1/2$. This is the probability that a face picked at random is blue. But let us calculate it again using the total probability theorem:

$$Pr(B_1) = Pr(BB \cap B_1) + Pr(BG \cap B_1) + Pr(GG \cap B_1)$$

$$= Pr(B_1|BB)Pr(BB) + Pr(B_1|BG)Pr(BG)$$

$$+ Pr(B_1|GG)Pr(GG)$$

$$= 1 \times \frac{1}{3} + \frac{1}{2} \times \frac{1}{3} + 0 \times \frac{1}{3} = \frac{1}{2}. \qquad (5.16)$$

Carefully check this calculation. Note that $Pr(B_1|BB) = 1$. Look at Fig. 5.9 and identify each of the regions in the figure and how they are related to the three terms in Equation (5.16).

We also know $Pr(BB) = 1/3$. Therefore, we get the result:

$$Pr(B_2|B_1) = \frac{\frac{1}{3}}{\frac{1}{2}} = \frac{2}{3}.$$

This is a new result. We have had three results so far. Which one do you think is the correct result?

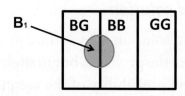

Fig. 5.9 Venn diagram for the three card problem.

First, let me discard the two quick arguments. In the first, we reach the result 1/3, arguing that the probability of the other side of the card being blue is equivalent to picking the BB card; therefore, the required probability is $Pr(BB) = 1/3$. In this calculation, we ignored the *condition* B_1. In the second quick argument, we argued that the event "second side blue" is equivalent to the event that the card on the table is *not* GG. Hence, it can be either BB or BG. Therefore, $Pr(BB) = 1/2$. This argument is faulty because we did not take into account the fact that we know that B_1 occurred.

We are left with the two solutions (I) and (II), which are based on the definition of the conditional probability.

While you ponder the difference between the two solutions, let me clarify an important point that might also be useful in the more difficult problem discussed in next section. This might also give you a hint as to which of the two solutions contains a slight fault.

Compare again Equations (5.13) and (5.15). The numerators are identical ($Pr(BB)$), but the denominators are different.

In the first solution, I treated the condition "given B_1" as if it was *equivalent* to $BB \cup BG$. But are these events really *equivalent*? Clearly, if we *know* B_1, then it follows that it cannot be GG. This means that it must be $BB \cup BG$. Up to this point, it is perfect. We write this as: $B_1 \subset BB \cup BG$.

We say that B_1 is *included* in $BB \cup BG$. "Included" means that if B_1 occurred, then it *follows* that $BB \cup BG$ occurred; a schematic Venn diagram of this is shown in Fig. 5.9. But to be *equivalent*, we must also show that the reverse is true, i.e. that $BB \cup BG$ is *included* in B_1. But this is not true. If we know that the event $BB \cup BG$ occurred, it *does not* follow that B_1 occurred. For instance, if $BB \cup BG$ occurred, it is still possible that G will show up on the table. We can state our intermediate conclusion as follows:

$$Pr(B_1|BB \cup BG) < 1$$
$$Pr(BB \cup BG|B_1) = 1.$$

Read these two statements and convince yourself that they are true. We can say that the "information" given in B_1 is "larger" than the information given in $BB \cup BG$. This formulation seems to contradict what you see in the schematic diagram in Fig. 5.9. The area B_1 is *smaller* than the area $BB \cup BG$. However, if I give you the *information* about the occurrence of B_1, it is "greater" in the sense that it is more detailed or more precise, and when I tell you that $BB \cup BG$ occurred, the area is larger, but the information I have is "smaller" (i.e. less precise). We shall return to discuss the size of information in Session 8. For now, think of the diagram in Fig. 5.9 as the locations that a dart hit. If I tell you that it hit region B_1, I am giving you more precise information than when I tell you that it hit $BB \cup BG$. From knowing that B_1 occurred, you *can* conclude the $BB \cup BG$ occurred, but from knowing that $BB \cup BG$ occurred, you *cannot* conclude that B_1 occurred.

I hope you see that in the first solution I did not use *all* of the information given, whereas in the second solution I used the *precise* information given.

Let us put it another way. If *all of the information* given in the problem is that $BB \cup BG$ occurred, then the first solution would have been correct. However, in the first solution (I), we used only *part* of the given information, not *all* of the information. In the second solution (II), we use *all* of the *given* information, and that is why the second solution is the correct one. Let us put it yet in another way. In the second solution, we wrote:

$$Pr(B_2|B_1) = \frac{Pr(B_2 \cap B_1)}{Pr(B_1)} = \frac{Pr(BB)}{Pr(B_1)} = \frac{2}{3}.$$

In the first solution, we wrote:

$$\frac{Pr(BB)}{Pr(BB \cup BG)} = \frac{1}{2}.$$

Now we can rewrite the second solution in terms of the first solution, but with a *correcting factor*:

$$\frac{Pr(BB)}{Pr(BB \cup BG)} = \frac{Pr(BB)}{Pr(B_1)} \frac{Pr(B_1)}{Pr(BB \cup BG)} = \frac{Pr(BB)}{Pr(B_1)} \times Factor.$$

All we did in the last equation is we started with the first solution, then we multiplied and divided by $Pr(B_1)$, and got the second solution with an additional factor. In words, we have:

$$\left(\begin{array}{c} Incorrect \\ solution \end{array} \right) = \left(\begin{array}{c} Correct \\ solution \end{array} \right) \times Factor.$$

Because $Pr(B_1) < Pr(BB \cup BG)$, the *factor* is less than unity. Therefore, the incorrect result of 1/2 is smaller than the correct result, which is 2/3.

I have taken the liberty of explaining this seemingly simple problem at great length. I will show you in the next problem that using the *correct information* could be critical in making a decision about life or death. In Session 8, we shall discuss the meaning of "larger" or "smaller" sizes of information.

5.4 The Monty Hall Problem

This is one of the most interesting examples in which our intuition fails us in assessing the correct probability.

Let's Make a Deal:
Monty Knows

Behind one of these doors is a car.
Behind each of the other two doors is a goat.
Click on the door that you think the car is behind.

Fig. 5.10 You can play the game at: http://math.ucsd.edu/~crypto/ Monty/monty.html. Or watch it at: http://www.youtube.com/watch?v= mhlc7peGlGg.

The Monty Hall problem gets its name from the TV game show "*Let's Make A Deal*," hosted by Monty Hall. The scenario is as follows: You are given the opportunity to select one out of three closed doors, behind one of which is a hidden prize. The other two doors hide "goats" (or some other such "non-prize"), or nothing at all. Once you have made your selection, Monty Hall will open one of the remaining doors, revealing that it does *not* contain the prize. He then asks you if you would like to switch your selection to the other unopened door, or stay with your original choice. Would you switch doors? You can play this game online (Fig. 5.10), or see a simulated game on YouTube.

The Monty Hall problem became popular after it was posed as a question from a reader's letter quoted in Marilyn vos Savant's "*Ask Marilyn*" column in *Parade* magazine in 1990. vos Savant's response was that the contestant should switch to the other door (vos Savant, 1990a).

Many of vos Savant's readers refused to believe that switching was beneficial despite her explanation. After the problem appeared in *Parade*, approximately 10,000 readers, including nearly 1000 with PhDs, wrote to the magazine, claiming that vos Savant was wrong. Even when given explanations, simulations and formal mathematical proofs, many people still do not accept that switching is the best strategy. Paul Erdos, one of the most prolific mathematicians in history, remained unconvinced until he was shown a computer simulation confirming the predicted result.

Here, we reformulate the problem using a more dramatic story about the three prisoners. The mathematical solution to the problem is identical to the Monty Hall problem.

5.5 The Three Prisoners' Problem: A Misuse of Probabilities That Can Cost You Your Life

This problem is important, interesting, challenging and enjoyable. Going through the solution should be a rewarding experience for the following reasons:

(1) It is a simple problem with a counter-intuitive solution.
(2) It illustrates the power of Bayes' theorem in solving probabilistic problems.
(3) It involves the concepts of "information" in making a "probabilistic decision," but in a very different sense from the way that this term is used in information theory (see Session 8).

This problem could be justifiably referred to as the prisoner dilemma. As we shall present it, it does involve a very serious dilemma. However, the term "prisoner dilemma" is already used in game theory for quite different problems. Therefore, we have chosen the title of "the three prisoners' problem." Read the problem carefully. Make notes on what is *given* and what the question is.

5.5.1 *The problem*

Three prisoners on death row, named A, B and C, await their execution, which is scheduled for the 1st of January. A few days before the said date, the King decides to free one of the prisoners. The King does not know any of those prisoners; he decides to throw a dice, the result of which will determine who shall be set free. On the six faces of this dice are inscribed the letters A, A, B, B, C and C. The dice is fair. The probability of each of the letters appearing is 1/3. The King then instructs the warden to free the prisoner bearing the name (i.e. the letter) that comes up on the face of the dice. He warns the warden not to divulge to the prisoners who the lucky one is until the actual day of the execution.

The facts that the King has decided to free one prisoner, that the prisoner was chosen at random with a 1/3 probability and that the warden has full knowledge about who is going to be freed but that he is not allowed to divulge it to anyone, are known to the prisoners. It is also common knowledge that the warden is impartial to the prisoners. Whenever he had to make a choice regarding the prisoners, he would toss a coin. A day before the execution, prisoner A approaches the warden and asks him: "I know that you are not allowed to tell me or anyone else who shall be freed tomorrow; however, I know that two of the three prisoners must be executed tomorrow. Therefore, either B or C must be executed. Please tell me which one of these two will be executed." Clearly, by revealing to prisoner A the identity of one of the prisoners who will be executed, the warden does not reveal the name of the one who will be freed. Therefore, he can answer prisoner A's query without defying the King's order. The warden tells prisoner A that prisoner B is going to be executed the next day.

The problem is this: Originally, prisoner A knew that the three prisoners have 1/3 probability of being freed. Now, he got some more *information*. He knows that B is doomed to die. Suppose A is given the opportunity to exchange names with C (note that the King decided to free the prisoner who, on the day of the execution, carries the *name* that appeared on the face of the die). Prisoner A's dilemma is this: Should he switch names with C (to render it more dramatic, we could ask: Should A offer C, say, $100 to swap names)? In probability terms, the question is the following: Initially, prior to asking the warden, A knows that the probability of his survival is 1/3. He also knows that the warden says only the truth, and that the warden is not biased (i.e. if he has to make a choice between two answers, he will choose between the two with equal probability). After receiving the additional information, the question is: What is the conditional probability of A's survival given the information supplied by the warden?

Intuitively, we feel that since A and C had the same initial probability of being freed, there should be the same probability for A and C to be freed after the warden informed A that B will be executed. Is this intuitive conclusion correct?

This problem can be easily solved using Bayes' theorem. The solution is counter-intuitive. The reader is urged to solve the problem before reading the solution below.

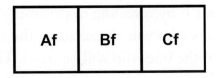

Fig. 5.11 The Venn diagram for the three events Af, Bf, and Cf.

5.5.2 *The solution to the three prisoners' problem*

We present here the solution to the three prisoners' problem in this section. Define the following three events (Fig. 5.11):

$$Af = \{\text{Prisoner A is going to be freed}\}, \quad Pr(Af) = \frac{1}{3}$$

$$Bf = \{\text{Prisoner B is going to be freed}\}, \quad Pr(Bf) = \frac{1}{3}$$

$$Cf = \{\text{Prisoner C is going to be freed}\}, \quad Pr(Cf) = \frac{1}{3}.$$

Define the following three events:

$$Ad = \{\text{Prisoner A is going to die}\}, \quad Pr(Ad) = \frac{2}{3}$$

$$Bd = \{\text{Prisoner B is going to die}\}, \quad Pr(Bd) = \frac{2}{3}$$

$$Cd = \{\text{Prisoner C is going to die}\}, \quad Pr(Cd) = \frac{2}{3}.$$

We also denote by $W(B)$ the following event:

$W(B) = \{$The warden points at B and says that he will be executed$\}$.

5.5.2.1 *First solution*

Prisoner A asks the question. As a result of the warden's answer, he *knows* that B will die. Therefore, the probability that A will be freed, given that B is going to die, is:

$$Pr(Af|Bd) = \frac{Pr(Bd|Af)Pr(Af)}{Pr(Bd)} = \frac{1 \times \frac{1}{3}}{\frac{2}{3}} = \frac{1}{2}. \qquad (5.17)$$

Read these equations and make notes of what each symbol means. Similarly:

$$Pr(Cf|Bd) = \frac{Pr(Cf \cap Bd)}{Pr(Bd)} = \frac{Pr(Bd|Cf)Pr(Cf)}{Pr(Bd)}$$

$$= \frac{1 \times \frac{1}{3}}{\frac{2}{3}} = \frac{1}{2}. \qquad (5.18)$$

According to this "solution," the conditional probabilities of either events *Af* or *Cf* are the same. This solution is intuitively appealing. The reasoning is simple: There were equal probabilities for *Af* and *Cf before* A asked the question, hence the probabilities remain equal *after* A asked the question.

This solution, although intuitively appealing, is wrong. The reason is that we have used the event *Bd* as the *given* condition; instead, we must use the event $W(B)$. If *given* the event *Bd* is the condition, then the probabilities in Equations (5.17) and (5.18) are correct. However, there is a difference between the "information" *Bd* and $W(B)$. Convince yourself that $W(B)$ is an event *contained* in *Bd*. In other words, if one knows that $W(B)$ occurred, then it follows that *Bd* is true. However, if *Bd* is true, it does not necessarily follow that $W(B)$ occurred. In terms of Venn diagrams (Fig. 5.12), we can see that the size (or the probability) of the event $W(B)$ (later we shall use the notation $(A \rightarrow B)$, see below) is *smaller* than that of *Bd*. Note that *Bd* is the complementary event to *Bf*, i.e. $Bd \cap Bf = \emptyset$ (Fig. 5.13).

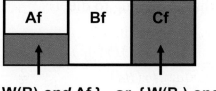

$$W(B)= \{W(B) \text{ and } Af\} \quad or \quad \{W(B) \text{ and } Cf\}]$$

Fig. 5.12 The event W(B) is a subset of the event Bd (which is Af and Cf).

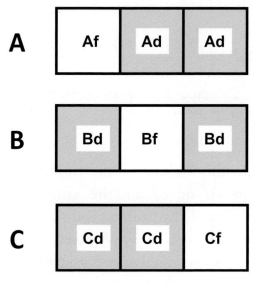

Fig. 5.13 The Venn diagram for the three events Af, and Ad, Bf, and Bd, Cf, and Cd.

We had $Pr(Bd) = 2/3$, but using the theorem of total probability, we can write the event $W(B)$ as:

$$W(B) = W(B) \cap Af \ \ or \ \ W(B) \cap Bf \ \ or \ \ W(B) \cap Cf \quad (5.19)$$

And so the probability of $W(B)$ is (see Fig. 5.12):

$$Pr(W(B)) = Pr(W(B)|Af)Pr(Af) + Pr(W(B)|Bf)Pr(Bf)$$
$$+ Pr(W(B)|Cf)Pr(Cf)$$
$$= \frac{1}{2} \times \frac{1}{3} + 0 \times \frac{1}{3} + 1 \times \frac{1}{3} = \frac{1}{2}. \quad (5.20)$$

The last equation means that the probability of the event $W(B)$ is the sum of the probabilities of the events on the right-hand side of Equation (5.19).

Note that in Equation (5.20), we capitalize on the fact that the warden is indifferent or unbiased towards A and C; if Af has occurred (i.e. prisoner A is going to be freed), then the warden can point at either B or C (see definition in the beginning of Section 5.5.2. We shall say for short, that the warden *points* at B). The choice he makes is with probability 1/2. On the other hand, if Cf occurred, then the warden does not have a choice but to point at B (as a result of A's question). Recognizing that the event $W(B)$ is smaller than Bd, in the sense that $W(B) \subset Bd$ (see Fig. 5.12), it follows that $W(B)$ contains more *information* than Bd. Therefore, we must use $W(B)$ instead of Bd in the solution to the problem.

5.5.2.2 *Second solution*

Instead of Equations (5.17) and (5.18), we write:

$$Pr(Af|W(B)) = \frac{Pr(W(B) \cap Af)}{Pr(W(B))} = \frac{Pr(W(B)|Af)Pr(Af)}{Pr(W(B))}$$

$$= \frac{\frac{1}{2} \times \frac{1}{3}}{\frac{1}{2} \times \frac{1}{3} + 0 \times \frac{1}{3} + 1 \times \frac{1}{3}} = \frac{1}{3}. \qquad (5.21)$$

Note again that $Pr(Af|W(B))$ is the probability that prisoner A will be freed given that the warden points at B.

$$Pr(Cf|W(B)) = \frac{Pr(W(B) \cap Cf)}{Pr(W(B))}$$

$$= \frac{1 \times \frac{1}{3}}{\frac{1}{2} \times \frac{1}{3} + 0 \times \frac{1}{3} + 1 \times \frac{1}{3}} = \frac{2}{3}. \qquad (5.22)$$

This means that by switching names, A doubles his chances of survival.

Thus, if we use the information contained in $W(B)$, we get a different result. Note that the amount of information contained in $W(B)$ *is larger* than that in Bd. We can express this as:

$$Pr(W(B)|Bd) = \frac{2}{3} < 1$$
$$Pr(Bd|W(B)) = 1.$$

Read these two equations carefully and make sure that you know what they mean. Compare these with the discussion in the case of the three cards problem in Section 5.3.

In the first solution of the problem, we used the *given* information Bd. This is less than the information that is *available* to A, which is $W(B)$. Therefore, the correct solution is based on Equations (5.21) and (5.22). In other words, by switching names, A can *double* his chances of survival.

5.5.2.3 *A more general but easier to solve problem*

The solution to the problem of the three prisoners is not easily accepted. It runs against our intuition, which tells us that if the probabilities of the two events *Af* and *Cf* were initially equal, then the equality of these probabilities must be maintained.

A simple generalization of the problem should convince the skeptic that the *information* given by the warden indeed changes the relative probabilities. This is an interesting example where a seemingly *more* difficult problem is *easier* to understand.

Consider the case of 100 prisoners named by the numbers "1," "2," ...,"100." It is known that only one prisoner is going to get freed, so the probability that a specific prisoner will be freed is 1/100. Suppose that prisoner "1" asked the warden, "Who, out of the remaining 99 prisoners, will be executed?"

The warden points to the following: "2", "3", ..., (he excludes "67"), ...,"100," i.e. the warden tells "1" the "names" of 98 prisoners, except "67," who will be executed.

Clearly, by switching names with "67," prisoner "1" increases his chances of survival from 1/100 to 99/100.

Note that, initially, prisoner "1" knew that he had a 1/100 chance of survival. He also knew that all of the remaining 99 prisoners together had 99/100 chances of survival. By acquiring the information on the 98 prisoners who will be executed, prisoner "1" still has a 1/100 chance of survival, but the chances of survival of anyone of the remaining 99 prisoners is now "concentrated" on one prisoner named "67." The latter has now a 99/100 chance of survival. Therefore, "1" should pay any price to switch names with "67." An illustration for ten prisoners is shown in Fig. 5.14. Note that in this and in the previous example, prisoner A had some initial *information* (on the probability of

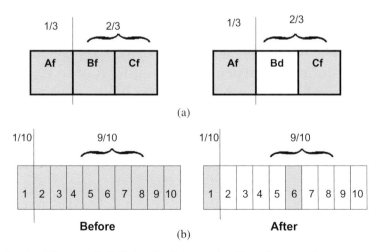

(a)

Before

(b)

After

Fig. 5.14 The probabilities before and after the warden answers the question. (a) The warden pointed at B to be executed. The probability 2/3 is now "concentrated" at Cf. (b) The warden pointed at: 2, 3, 4, 5, 7, 8, 9, 10 to be executed. executed. The probability 9/10 is now "concentrated" at "6."

being freed) and also had received some additional *information*. In Session 8, we shall discuss a *measure* of the information obtained by prisoner A.

5.5.2.4 *An apparent paradox*

A student who followed the arguments of the solution to this problem asked: Suppose that C asks the *same* question and the warden answers with the same answer, i.e. the warden indicates that B is to be executed. Therefore, given $W(B)$, we should conclude, by the same arguments as above, that it is in C's advantage to switch names with A. But we have just concluded that it is in A's advantage to switch names with C.

It sounds as if we reach two conflicting conclusions, given the same information $W(B)$. If A asks the question of the warden, then $Pr(Af|W(B)) = \frac{1}{3}$ and $Pr(Cf|W(B)) = \frac{2}{3}$. But if C asks the question of the warden, then $Pr(Cf|W(B)) = \frac{1}{3}$ and $Pr(Af|W(B)) = \frac{2}{3}$. How did we arrive at different conclusions based on the *same* information given to either A or to C?

The apparent paradox is a result of our reference to $W(B)$ as the *same* information, when given to either A or to C.

It is true that the answer given by the warden "B will be executed" is *literally* the *same* answer when given to A or when given to C. It is the *same* in the sense that it uses the same *words*. But the message it carries is different when it is addressed to A or to C.[3]

In order to remove the apparent paradox, we should redefine $W(B)$ more precisely; instead of the definition of $W(B)$ in Equation (5.19), we define the event:

$$W(A \rightarrow B) = \{\text{The warden points at } B \text{ to be doomed}$$
$$\text{as a result of A's question}\}. \qquad (5.23)$$

The arrow in (5.23) means that pointing at B follows from A's question.

Similarly:

$$W(C \to B) = \{\text{The warden points at B to be doomed}$$
$$\text{as a result of C's question}\}. \qquad (5.24)$$

The arrow in (5.24) means that pointing at B follows from C's question.

Clearly, A and C ask different questions. Prisoner A asks about the remaining prisoners B or C, while C asks about A or B.

Now it is clear that the two events (5.23) and (5.24) are not the same. The reason is that the warden *knows* who is to be freed; therefore, when answering either prisoner A or C, he might or might not point to the same prisoner. In terms of Venn diagrams, we can write the events (5.23) and (5.24) as (Fig. 5.15):

$$W(A \to B) = [W(A \to B) \cap Af] \text{ or } [W(A \to B) \cap Bf]$$
$$\text{or } [W(A \to B) \cap Cf]$$

The corresponding probability is:

$$Pr(W(A \to B)) = \frac{1}{2} \times \frac{1}{3} + 0 \times \frac{1}{3} + 1 \times \frac{1}{3} = \frac{1}{2}. \qquad (5.25)$$

Similarly,

$$W(C \to B) =$$
$$W(C \to B) \cap Af \text{ or } W(C \to B) \cap Bf \text{ or } W(C \to B) \cap Cf$$

The corresponding probability is:

$$Pr(W(C \to B)) = 1 \times \frac{1}{3} + 0 \times \frac{1}{3} + \frac{1}{2} \times \frac{1}{3} = \frac{1}{2}. \qquad (5.26)$$

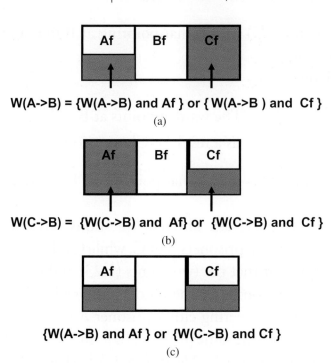

$$W(A\text{->}B) = \{W(A\text{->}B) \text{ and } Af\} \text{ or } \{W(A\text{->}B) \text{ and } Cf\}$$

(a)

$$W(C\text{->}B) = \{W(C\text{->}B) \text{ and } Af\} \text{ or } \{W(C\text{->}B) \text{ and } Cf\}$$

(b)

$$\{W(A\text{->}B) \text{ and } Af\} \text{ or } \{W(C\text{->}B) \text{ and } Cf\}$$

(c)

Fig. 5.15 Various Venn diagrams for the solution of the three prisoners problem.

Note that the events $W(A \rightarrow B)$ and $W(C \rightarrow B)$ are *different* (Fig. 5.15) but their probabilities (size of the area in the Venn diagram) are the same.

Since the warden *knows* who is going to be freed, the information given to A is different from the information given to C. Therefore, the conditional probabilities will be different depending on who receives the information. However, it is true that the warden answers to both prisoners A and C by using the *same words*, i.e. "B is doomed to die," but the significance of this information is different if given to A as a result of A's question or as a result of C's question. The solution we reached above is that if $W(A \rightarrow B)$ is true, then it is in A's advantage to exchange names with C. If, on the other hand, C receives the *same* information, i.e. if C knows $W(A \rightarrow B)$, then it is still in A's advantage to exchange names. It is only when C asks the warden and gets the *different* information $W(C \to B)$ that it is in C's advantage to exchange names with A. Thus, in this problem, we have four different conditional probabilities:

(1) Prisoner A asks and receives the information $W(A \to B)$:

$$Pr(Af|W(A \to B)) = \frac{Pr(W(A \to B)|Af)Pr(Af)}{Pr(W(A \to B))}$$

$$= \frac{\frac{1}{2} \times \frac{1}{3}}{\frac{1}{2}} = \frac{1}{3}. \tag{5.27}$$

(2) Prisoner A asks, and C receives the information $W(A \to B)$:

$$Pr(Cf|W(A \to B)) = \frac{Pr(W(A \to B)|Cf)Pr(Cf)}{Pr(W(A \to B))} = \frac{2}{3} \tag{5.28}$$

(3) Prisoner C asks and receives the information $W(C \to B)$:

$$Pr(Cf|W(C \to B)) = \frac{Pr(W(C \to B)|Cf)Pr(Cf)}{Pr(W(C \to B))}$$

$$= \frac{\frac{1}{2} \times \frac{1}{3}}{\frac{1}{2}} = \frac{1}{3}. \tag{5.29}$$

(4) Prisoner C asks and receives the information $W(C \to B)$:

$$Pr(Af|W(C \to B)) = \frac{Pr(W(C \to B)|Af)Pr(Af)}{Pr(W(C \to B))}$$

$$= \frac{1 \times \frac{1}{3}}{\frac{1}{2}} = \frac{2}{3}. \tag{5.30}$$

Thus, we see that if either A or C receives the *same* information, then the conclusion is the same and there is no paradox.

To complete the list of possible cases, we should add two more cases:

(5) No one asks any questions. In this case, the probabilities are equal, i.e.:

$$Pr(Af) = Pr(Bf) = Pr(Cf) = \frac{1}{3}.$$

(6) Both A and C ask the same question and the warden points at B. Both A and C know the two answers. In this case, the known information is $W(A \to B) \cap W(C \to B)$ and the probabilities are:

$$Pr(Af|W(A \to B) \cap W(C \to B))$$
$$= \frac{Pr(W(A \to B) \cap W(C \to B)|(Af)Pr(Af))}{Pr(W(A \to B) \cap W(C \to B))}$$
$$= \frac{\frac{1}{2} \times \frac{1}{3}}{\frac{1}{3}} = \frac{1}{2} \qquad (5.31)$$

and similarly:

$$Pr(Cf|W(A \to B) \cap W(C \to B)) = \frac{1}{2}. \qquad (5.32)$$

Note that the size (probability) of the intersection event is 1/3 (see Fig. 5.15) In this case, both A and C receive the *same* information, hence, no one will have an advantage if they switch their names.

Exercise: In the solution given in (5.21) and (5.22), we have assumed that the warden is completely indifferent in the sense that if he knows *Af*, then he chooses to point at either B or C with equal probabilities. Equivalently, he tosses a fair coin in order to make his decision. Suppose that prisoner A asks the same question and he knows that the coin the warden uses (if indeed he uses one) is unbalanced. Therefore, if the warden needs to make a decision between B and C (i.e. when he knows Af), he tosses the coin with probabilities.

$$Pr(W(A \rightarrow B)|Af) = x$$
$$Pr(W(A \rightarrow C)|Af) = 1 - x.$$

How does this new information affect A's decision on whether to switch or not to switch names?[4]

5.6 Re-appraisal of Probabilities

Probabilities are sometimes used as a measure of the extent of the truth of some statement, proposition, etc. We write $Pr(A)$ to mean the probability that A is true. Now suppose that we have new information I that either supports or does not support the notion that A is true. We write this using the conditional probability:

$$Pr(A|I) = \frac{Pr(A \cap I)}{Pr(I)} = \frac{Pr(I|A)Pr(A)}{Pr(I)}$$
$$= Pr(A)\frac{Pr(I|A)}{Pr(I)} = Pr(A)g(A, I).$$

On the right-hand side, I have written the original or the *a priori* probability of A, then multiplied this by a quotient to get the new, modified probability of A given the information I. The modifier is simply the correlation between A and I.

The factor $g(A, I)$ can either diminish or enhance the probability of A. Next, suppose that we get new information J. The new re-appraised probability is now the probability that A

is true given both information I and J:

$$Pr(A|I \cap J) = \frac{Pr(A \cap (I \cap J))}{Pr(I \cap J)} = \frac{Pr(A)Pr(I \cap J|A)}{Pr(I \cap J)}$$
$$= Pr(A)g(A, I \cap J).$$

Now, we have a new appraisal of the probability of A. The modification is measured by the correlation between $I \cap J$ (given both I and J) and A.

The above formulation is very abstract, but in fact what is written above is actually how we think every day. I shall give you two quick examples, then a lengthier one with a counter-intuitive result.

(1) Suppose we find some pieces of scrolls in an excavation site. We initially assess that the probability of the proposition A, that these scrolls were written by King Solomon, is $Pr(A) = 0.6$. Next, we find at the same site a golden ring with the initials KS inscribed on it. This new information (I) suggests that the ring belonged to King Solomon, and enhances our belief that the scrolls that were found were indeed written by him. We therefore re-appraise our probability of the proposition A as:

$$Pr(A|I) = 0.6 \times g(A, I) = 0.6 \times 1.5 = 0.9$$

where $g(A, I)$ is the correlation between the proposition A and the new information I, which, in this instance, has been given the value of 1.5.

(2) You read in the newspaper that a man found an old booklet in his attic with some musical notes signed by W. A. Mozart. The musical piece was characteristic of Mozart's style. Therefore, the man claimed that the proposition (A) that this piece was composed by Mozart himself, but was never

published before, is almost certainly true, say $Pr(A) = 0.9$. The man offers the booklet to Mozart's museum, which rewarded him with an undisclosed amount of money.

A few years later, the paper on which the composition was written was examined by a team of scientists. They found that the paper used was not produced before the 20th century, and therefore the alleged composition could not have been written and composed by Mozart himself. The new information (I) does not support the initial proposition. The re-appraised probability that A is true was now:

$$Pr(A|I) = Pr(A)g(A, I) = 0.9 \times 0.01 = 0.009.$$

In the first example, the new information supported the proposition A. In the second example, it did not support the initial proposition.

(3) Another business offer

My friend Naivic Inventoric told me that he was about to invent a super-problem-solver-device (SPSD), and that very soon he might possibly earn millions. He suggested that I invest some money in the project, promising a huge profit when the product was launched in the market. I asked him what the new product's probability of success was. He thought for a while and said, "To be honest, I do not know. However, a fair guess is that it should have a 50% chance of success or not." This seemed fair and honest, but I declined the offer.

After a few days, my friend came back and informed me that a survey was conducted and that 70% of the respondents said that they *would buy* the SPSD as soon as it is launched. The new information made him re-appraise his original probability of success from 50% to 70%. I was still hesitant. A few days later, my friend came back again with new information. Apparently, a new survey was conducted and the results indicated that almost 95% of the respondents said that they were very eager to have a device that would solve all of their problems. He said that, based on this new information, he had re-appraised the probability of success of the SPSD to over 90%. "Perhaps even higher," he added, "but at least 90%." I replied, "This is indeed impressive. Let me think about it, I need to consult my book dealing with such problems, as well as with my readers."

Would you, the reader, invest in the development of the SPSD, given such a high probability of its success?

I hope that you have detected the pitfalls in this proposal. I know that my friend is innocent, and he did not want to drag me into this dubious investment. I believe he was convinced that his SPSD had a huge chance of success. Unfortunately, even his initial assessment of the probability of success was doubtful. Furthermore, the (presumably true) fact that 70% of the respondents said that they *would buy* the SPSD as soon as it was launched does not lend any support to claim that such a SPSD device would soon be invented. Likewise, the (presumably true) fact that 95% of the respondents said that they *were eager to have such a device* (I myself was also eager to have one) does not say anything regarding the claim that such a device would ever be invented and introduced in the market.

In conclusion, the whole argumentation is false. Perhaps my friend was naïve and he believed that the two surveys enhanced his chances of being rich, but I could not accept this offer.

The story told above is fictitious. You might be thinking that I am simply kidding here. Unfortunately, such deceptive arguments, either intentionally or unintentionally, are often made by politicians who are soliciting for your votes, or by drug companies who want to sell you a new drug that will make you feel 20 years younger.

5.7 An Abuse of Probability and Bayes' Theorem

An outright abuse of probabilistic arguments (involving Bayes' theorem) may be found in Stephen Unwin's book "*The probability of God: A simple calculation that proves the Ultimate Truth*" (2004).

First, recall that scientific probabilities are assigned to events or to outcomes of an experiment, not to objects. In probability theory, one never talks about the "probability of a table," nor of the "probability of the moon" or of the "probability of Kukuriku." Therefore, the title of Unwin's book is misleading. On the book's first page, Unwin writes: "I use the term probability in its strict *mathematical sense* and not in the fuzzy, ambiguous way it can be used in common language." That claim is an outright deception. The book deals *exclusively* with "the fuzzy, ambiguous" meaning of probability. The proposition "G" as defined by Unwin as "God exists," which is neither an event nor an outcome of an experiment.

On page 15 of his book, Unwin is careful to note that the proposition "I did not have sexual relations with that woman" cannot be judged as true or false unless one *defines* what is meant by "sexual relations." This is correct. However, there is an abounding discussion on the probability of the proposition "God exists" in the entire book without ever defining who, or what, "God" is. This renders the entire contents of the book meaningless.

On page 58, Unwin writes: "Here, I think that the expression of complete ignorance [of the truth of proposition "G"] is a good case of the 50–50 argument." It is not! He continues: "This is a perfect, unbiased expression of agnosticism." It is not! Clearly, Unwin draws from the symmetry argument applied to the two outcomes of throwing a coin. In the case of proposition "G," it is not clear what the experiment is, nor what the outcomes are. Therefore, the assignment of 50% probability to proposition "G" is totally arbitrary.

The rest of the book is a futile exercise in using "evidentiary areas" in connection with Bayes' theorem to reassess the probability of proposition "G."

Note that I am not arguing here that the proposition "God exists" is not true. What I am saying here is that one cannot use Bayes' theorem to prove that the proposition "G" is true or false.

Using the same language and the same kind or reasoning, one can prove anything one wants, and get any probability one wants. Specifically, one can "prove" that the important proposition "K," the probability of "the existence of Kukuriku," is 95%. Here is the proof:

First, assume that the *a priori* probability of Kukuriku is 0.5. Is this a fair assumption? In Unwin's words: "This is a perfect, unbiased expression of agnosticism."

Next, we re-appraise this probability. We know that many children utter the sound "Kukuriku." Therefore, this information *supports* the proposition "K." The new re-appraised probability is now 0.8. Second, we sometimes hear at daybreak, perhaps twice or even thrice, a sound emanating from the farm that reminds us of the existence of "Kukuriku." This certainly supports our assertion that Kukuriku exists, i.e. this new "evidence" supports the proposition "K." Now, our re-appraised probability is 0.95 — an almost certainty!

You see, I have just proved that Kukuriku exists with 0.95 or 95% probability. I hope you were convinced.

You might think that this is a joke, but no, it is not a joke, not even a bad joke. It is sheer nonsense, disguised in scientific language.

Finally, the subtitle of the book is also meaningless, unless one defines what the "Ultimate Truth" is. Once we know what the "Ultimate Truth" is, we shall not need to prove it!

5.8 The Solution to the Urn Problem

There are four balls in an urn, two white (W) and two black (B). We asked two questions:

(1) What is the probability of drawing a white ball in the second draw (call this event W_2), given that in the first draw we picked a white ball (call this event W_1), and it was not returned to the urn?

(2) What is the probability that we drew a white ball in the first draw, given that the second draw was white?

The answer to the first question is straightforward. We write it as:

$$Pr(W_2|W_1) = \frac{1}{3}.$$

The reasoning is simple. After we drew a white ball in the first draw, we are left with three balls in the urn — one white and two black. Therefore, the conditional probability of ($W_2|W_1$) is 1/3.

The answer to the second question is also simple. The question is, "what is the conditional probability of W_1 given W_2?" We write this as:

$$Pr(W_1|W_2) = ?$$

From the definition of the conditional probability, we have:

$$Pr(W_1|W_2) = \frac{Pr(W_1 \cap W_2)}{Pr(W_2)}.$$

$Pr(W_2)$ is the (unconditional) probability of drawing a white ball on the second draw. This can be calculated from the *total probability theorem* (or just by an argument of symmetry):

$$Pr(W_2) = Pr(W_2|B_1)Pr(B_1) + Pr(W_2|W_1)Pr(W_1)$$
$$= \frac{2}{3} \times \frac{1}{2} + \frac{1}{3} \times \frac{1}{2} = \frac{1}{2}.$$

The probability that a white ball will be drawn on both the first and the second draws is:

$$Pr(W_1 \cap W_2) = \frac{1}{2} \times \frac{1}{3} = \frac{1}{6}.$$

Therefore, the required conditional probability is:

$$Pr(W_1|W_2) = \frac{\frac{1}{6}}{\frac{1}{2}} = \frac{1}{3}.$$

The psychological reason that the second question seems more difficult (some even say it is impossible) is that one tends to confuse conditional probability with cause and effect: How can an event in the present affect the probability of an event in the past? In other words, we associate the conditional probability with the arrow of time. (Note that, in this particular case, the event W_1 came first, and W_2 came second. However, in general, conditional probability does not mean that the "condition" precedes the event that we are interested in.)

One way to overcome this illusion is to reformulate the same problem in a slightly different way. We pick a ball and hide it. Then we pick another ball and see that it is white (W_2). Now we ask: What is the probability that the hidden ball is white (W_1), given that we know that the second ball that was drawn was white? In this formulation, we have "reversed" the order of events in time; we "sent" the first event (W_1) from the "past" into the "future."

If you are not convinced, consider the following simpler exercise. There are only two balls in the urn — one white and one black. The questions are the same as above. The answers in this case are immediate:

$$Pr(W_2|W_1) = 0$$
$$Pr(W_1|W_2) = 0.$$

If you like to prove a general mathematical theorem, consider the following exercise: There are N white and N black balls in an urn. We ask exactly the same questions as above. Calculate $Pr(W_2|W_1)$ and $Pr(W_1|W_2)$.[5]

5.9 Conclusion

This was a long session. We have used some mathematics. However, if you examine all of the mathematics used here, you will find that, in effect, we used only four basic arithmetical operations. We used the term "theorem" twice — in the "total probability theorem" and in "Bayes' theorem." Both of these use only the four elementary arithmetic operations plus some logical arguments, and nothing more.

I hope that the selected examples presented in this session have convinced you of the importance of Bayes' theorem. It is important not only in solving recreational problems (e.g. the Monty Hall problem), but also far more serious problems (e.g. false-positive results). I also hope that by carefully studying Bayes' theorem, you have acquired the tools to critically examine some assertions made in the literature that are "fashioned" in a scientific language but are nothing more than sheer nonsense.

Session 6

Average, Variance and Random Variables

Before we define the concept of an average, let me remind you that you have already used this concept in previous sessions. So I presume that you know what an average means. Let us start with a few stories involving averages.

6.1 To Buckle or Not to Buckle

Some years ago, before it was even mandatory to use seatbelts in vehicles, I tried to convince my friend Gary that it is safer to buckle up even if the law did not require it.

Gary was adamantly against it. His argument was compelling. He told me that he had a friend who had been using his seatbelt when he was involved in a horrific accident that engulfed the car in flames. Although he was not hurt by the impact of the collision, he could not disengage himself from the seatbelt, and as a result he was burned alive. His wife, who was seated beside him, was not buckled up, but was thrown out of the car as the car tumbled and rolled over. She nevertheless survived and only sustained minor injuries.

The moral, Gary snapped at me, was that he would never use seatbelts even if the law required it.

I realized that his reasoning was logical, as well as supported by the traumatic accident he had told to me. I had no chance of convincing him to use a seatbelt. However, my conviction remains steadfast. Can you explain why?[1]

6.2 Can an Average Grade be Higher Than the Highest Grade?

The average grade of all the students in the Average-State University (ASU) in 1970 was 83.4 (100 is the maximal grade). This is quite good compared with the grades of all of the students in the country. In the years that followed, the average grade of the students of ASU declined each decade, starting at 83.4 in 1970 and dropping down to 82.1 in 1980, then dropping further to 79.1 in 1990. Finally, in 2000, the ASU's local newspaper proudly announced that the grades of *all* of the students of ASU were *above the average*.

Is this good news for a change?[2]

6.3 How Can One Increase the Average IQ of the Professors in Two Universities?

In another publication, it was reported that in Highiq State University (HSU), the average IQ of university professors was 130. In that same publication, it was also reported that the average IQ of university professors at the Lowiq State University (LSU), which was located in the next town, was only 80 (please do not take any offense if you belong to LSU. The numbers I quoted here are purely fictitious).

Two years ago, a professor from the HSU took a position at LSU. At the end of the academic year, it was found that as a result of this transfer, the *average* IQ of the professors of each of the two universities had increased.

Could this be possible? The answer is "yes," and you can easily find an example.[3]

Does the increase of the average IQ of the professors of each university imply that the average IQ of *all of the professors* in the two universities had also increased? In this case, this is not possible. Can you think why?

6.4 Average Speed and the Average of Two Speeds

You drive from Jerusalem to Tel Aviv at a constant speed of 40 km/h. You drive back from Tel Aviv to Jerusalem at constant speed of 100 km/h. What is the average speed in the round trip to Tel Aviv and back?[4]

If you are having difficulty in solving this problem, try the following "easier" one, but with a surprising result.

On the way from Jerusalem to Tel Aviv, you ride on a donkey. The speed of the donkey is v. On the way back, you fly at nearly the speed of light; call this speed c. You know that $c \gg v$. What is the average speed of the round trip, and what is the average of the two speeds? A rough estimate will be accepted.[5]

If you did the two exercises above, you might want to try the following related problem.

You are told that there is no wind today and the velocity of a plane flying from Tel Aviv to New York and back is constant at v. Tomorrow, there will be a strong wind from west to east, such that the ground speed on the way to New York will be $v - w$ (w being the constant speed of the wind). On the way back, the plane will have a back-wind and its speed will be $v + w$. Clearly, the average of the two speeds is simply v independent of w.

Assuming that you want to minimize the flying time in a round trip to New York and back, on which day will you choose to fly — today or tomorrow?[6]

I have brought up these little stories in the opening of this session to show you that although we did not define the concept of an average, you already have a sense of what it is, and you already know how to calculate it in some simple cases.

Let us *formalize* the definition of the average for a simple example, then generalize it.

Suppose you measure the temperature at 12 noon every day. You get a list of numbers (all in degrees Celsius)
Day: 1, 2, 3, 4, 5, 6, 7, ...
Temperature: 17, 18, 17, 20, 22, 24, 24, 23, ..., 27, 29, 27.

You collected these data for a year. You have 365 numbers in the list. What is the average temperature in the specific location where the measurements were taken?

This is very simple. You sum up all the temperatures you have in the list and divide it by 365. It is convenient to use the following notation. We denote by $T(i)$ the temperature measured on day i, where i is an index running from $i = 1$ (the first day) to $i = 365$ (the last day). We now write the average temperature as

$$\bar{T} = \langle T \rangle = \frac{\textit{sum over all } T(i)}{365} = \frac{\sum_{i=1}^{365} T(i)}{365}$$

Either of the notations $\bar{T}$ or $\langle T \rangle$ are common for an average. The symbol $\sum_{i=1}^{365}$ is a shorthand notation for "the sum over all indices i, from $i = 1$ to $i = 365$."

We can also plot the recorded data from over the year, or over one month, and draw the average line (Fig. 6.1). We see that the average is somewhere between the largest values and the lowest values of the temperature. Figure 6.2 shows the same plot at different places of the world: Cairo, Rome and Moscow.

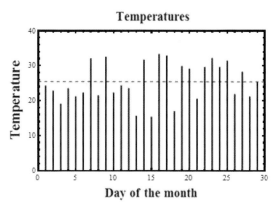

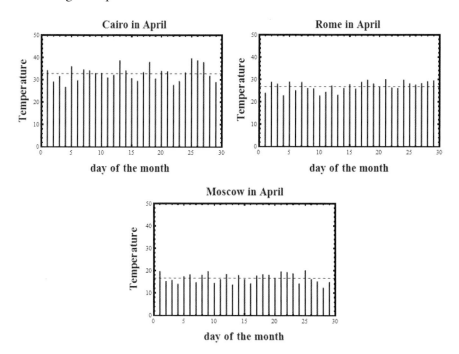

Fig. 6.1 Temperatures at various days in one month. Dashed line shows the average temperature in that month.

Fig. 6.2 Temperatures at three cities: Cairo, Rome and Moscow.

One can instantly notice that the average temperature in Rome is lower compared with Cairo, and the average temperature in Moscow is even lower than that of Rome. Note that on some days the temperature in Rome might be higher than in Cairo and lower than in Moscow, but the overall picture is clear — the average temperature is highest in Cairo and lowest in Moscow.

Let us take another example. You measure the heights of all of the people in a city and you get a list of numbers. Let us denote the height of the person with the index i by $h(i)$. Then, you can write a table of the form:

Person's index: 1   2   3   4 ...
Person's height: 1.70   1.67   1.80   1.71 ... .

The average height of the city's inhabitants (assume that there are 1000 people and their height is measured in cm) is written again as:

$$\bar{h} = \langle h \rangle = \frac{1}{1000} \sum_{i=1}^{1000} h(i).$$

You can also plot the result that shows the values of $h(i)$ as a function of the index i, as well as the average height in that population. A typical plot might look like the one in Fig. 6.3.

Children often tell the joke about the man who drowned in the lake. He heard that the *average* depth of the lake was 50 cm,

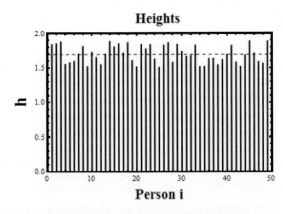

Fig. 6.3   Heights of fifty persons.

and so he walked into the lake even though he did not know how to swim. (Explain the confusion between average and the most probable.)

There is another way to plot the same data. Suppose that instead of listing in a table all of the persons and their corresponding heights individually, as we did in Fig. 6.3, we list all of the *possible heights*, and how many persons have each particular height. For example, if the lowest height recorded is 156 cm, and the highest is 193 cm, then we might have a table of the form:

Heights:	156	160	165	170	175	180	185	190	193
Number of persons:	1	23	100	320	500	50	4	2	1

The second row has the number of person with the height as listed in the first row.

In this case, you write the average exactly as before. We first collect all of the persons in the city into groups such that there are $N(h)$ persons having height h and sum over all possible heights:

$$\langle h \rangle = \frac{1}{1000} \sum_{all\ possible\ heights\ h} N(h)h.$$

The number $\frac{N(h)}{1000}$ is the fraction of the persons in the population having height h. If the population is very large, we can also interpret this fraction as the probability of a person being picked at random from the population and having height h, i.e. we define $Pr(h) = \frac{N(h)}{1000}$, or in the more general case we divide $N(h)$ by the total number of persons in the city. Thus, we arrive at the more convenient definition of the average height of the entire population in the city:

$$\langle h \rangle = \sum_{h} Pr(h)h.$$

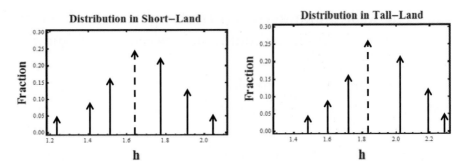

Fig. 6.4   Distribution of heights.

The sum is over all of the possible values of *h*.

It is also convenient to plot a graph of *Pr*(*h*) as a function of *h*, as in Fig. 6.4, rather than the graph of *h* as a function of the index *i*, as we did before in Fig. 6.3. In this case, we say that we have a plot of the *distribution* of heights in that particular city. You might collect data on the heights of people in different cities, for instance in Tall-Land and Short-Land, and you will get two plots that will look similar, but one may be shifted towards lower values of *h* than the second. Furthermore, the average height in Tall-Land is larger than the average height in Short-Land. However, take note that the shapes of the two plots are very similar. We shall discuss the origin of this shape in the next session. Note also that the most probable height might or might not coincide with the average height.

Figure 6.5 shows the distribution of speeds of argon atoms at different temperatures. This distribution was calculated theoretically. It is called the Maxwell–Boltzmann distribution. It is one of the greatest achievements of science in the late 19th century. Not only did the scientists find the distribution of speeds, but they also related this distribution to temperature.[7]

For our purposes, we need not know how this distribution was obtained. You can assume that a tiny speed detector was

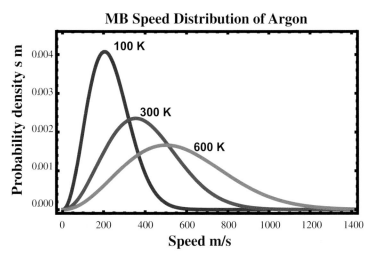

Fig. 6.5 Distribution of speeds (in meters per second) of argon at different temperatures.

positioned in the corner of the room and measured the speed of each molecule and plotting results on a graph as shown above in Fig. 6.5.

There are several parameters describing or characterizing the distribution curve. The simplest is the average or the mean value. In mathematics, this is also referred to as the *expected value*. (Note, however, that this term might be misleading. The "expected" value is not necessarily one of the actual measured values, nor even the most probable value. It means that, on average, over many measurements, you will get the expected value, i.e. the average.)

6.5 Standard Deviation

Another quantity that characterizes a distribution is the standard deviation. If you look at the three curves of the speed distribution in Fig. 6.5, you will see that at a very low temperature, the distribution is sharper, and most of the area under the curve (blue) is located at the lower speeds. On the other hand, at higher temperatures, the distribution is spread over much larger values of the speed, and most of the area under the curve (red) is shifted to the right, i.e. to higher temperatures. In fact, physicists have found that the width of the distribution of speeds of molecules is related to the temperature of the system containing the atoms.

How is the standard deviation measured? Look at the two curves in Fig. 6.6. These two curves have the same average. However, they are very much different from each other. The first (a) looks more concentrated near the average value, while the second (b) is much more spread over larger values of the temperatures. We wish to quantify this difference between these curves. As a measure of this quantity, we might consider the distance of each point on the curve from the average value. Denote this distance by $(T - \bar{T})$. What is the average of the quantity $(T - \bar{T})$? This is the *average distance* from the *average value* of the temperature. As usual, this average is defined by:

$$\sum_{T} Pr(T)(T - \bar{T}).$$

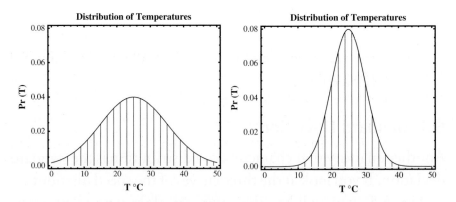

Fig. 6.6 Distribution of temperatures.

To calculate the average, we take the quantity $(T - \bar{T})$ (we assume that T takes only discrete values, not continuous values as shown in Fig. 6.6), then multiply this by the probability of finding the value of that temperature T, $Pr(T)$, then sum over all possible values of T. Note that, in principle, we should sum over T from zero to infinity (well, not strictly infinity, but very large values instead). However, in practice, this sum is only over a finite range of temperatures. We may neglect values of T that have negligible probability of occurrence.

If we evaluate this average quantity, we shall find that the result is zero. Can you prove this?[8]

What we need is a measure of the *absolute* distances from the average. One possibility is to take all of the absolute distances. This is denoted by $|T - \bar{T}|$, and then we take the average values of this quantity.

The absolute value of a number X, denoted $|X|$ is defined as follows: $|X| = X$ when X is positive and $|X| = -X$ when X is negative. Therefore, $|X|$ is always a positive number. The average of $|T - \bar{T}|$ is defined as:

$$\sum_T Pr(T)|(T - \bar{T})$$

This quantity, by definition of the absolute value, must be a positive quantity. It is a legitimate measure of the average spread about the average value. However, mathematicians have found it more convenient to take the square of the distances, i.e. $(T - \bar{T})^2$, which is again positive. We take the average value of this quantity, which is defined as:

$$\sum_T Pr(T)(T - \bar{T})^2$$

Note that we used both the upper bar and the brackets ⟨ ⟩ for the average.

The standard deviation is defined as the square root of this quantity. It is this standard deviation of the speed distribution of molecules that was found to be proportional to the temperature of a macroscopic body. This is not a trivial connection between two very different concepts. On the one hand, we have temperature, which we can sense with the tip of our finger or measure with a thermometer. Nothing in this sensing of temperature indicates that the temperature has anything to do with the motion of the molecules. On the other hand, we know what the speed or the velocity is of a body. We can observe a hot ball moving slowly or even not moving at all, and a very cold ball moving very fast. Nothing in these two observations relates the speed of the entire ball with its temperature. Yet in the *microscopic world*, the motions of the atoms and the molecules do determine the temperature as we sense or measure it in the macroscopic world.

There is one more very important point from physics regarding standard deviation. This is the relationship between the standard deviation of the energies of the molecules and the heat capacity of the (macroscopic) body. You will not need this quantity in your study of the theory of probability. I bring it up here to show you that standard deviations are very useful quantities. They appear in physics in sometimes unexpected contexts. Today, unlike towards the end of the 19th century, almost all of physics is based on probabilistic concepts. In particular, thermodynamics and statistical thermodynamics are built upon probabilities, averages and standard deviations.

Before we continue, let me digress to tell you what heat capacity is and why the heat capacity of water is important to life. The heat capacity (or specific heat) of a substance is defined as the amount of heat you have to add to, say, 1 cm^3 of a substance in order to raise its temperature by 1°C. In short, a substance with a high heat capacity will "resist" an increase in its temperature more than a substance with a low heat capacity (for the same quantity of the substance and the same amount of heat).

Water has an unusually high heat capacity. Most of the chemical reactions in our body are carried out in aqueous environments. Some of these reactions produce heat, which causes an increase in temperature. Water can absorb the heat produced by these chemical reactions with a minimal increase in its temperature. Thus, in effect, water serves as a buffer in order to maintain, as far as possible, a constant temperature in the body.

6.6 Random Variables

Random variables are very important in probability theory. You shall not need this concept at this stage of learning the basics of the theory. However, I shall indicate here one reason why one needs to introduce this concept.

Until this point in this session, we have discussed average quantities of heights, temperatures, velocities, etc. You had no difficulty in understanding the *meaning* of the various averages, as well as the method of computing the average.

Here is a quick test that I am right: You are given a fair dice. You throw it many times. What is the average value you would expect to get? What would be the distribution of the values?

I am sure you know the answer to these questions.

Now you are given a *fair coin* and you throw it many times. What would be the average value? If you have difficulties with coins, let us go back to the fair dice. However, this dice does not have any dots on its faces. Instead, each face is colored with

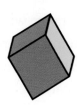

Fig. 6.7 A dice with colored faces.

a different color, say blue, red, green, yellow, brown and white (Fig. 6.7).

What is the average result you would expect to get?

Unlike for the first dice, where you can easily calculate the average outcome, here the average outcome is not defined.

In all of the average quantities we calculated above, we collected *numerical* data; heights of persons, temperatures in some city at different days, etc. But in general, an outcome of an experiment might not have a numerical value, like the outcome of a coin, the color of the dice face or the shape of the object you pick from an urn. In all of these experiments, the average is not defined. In some cases, however, it is convenient to assign numerical values to the various outcomes, whatever these outcomes are; e.g. numbers, colors, shapes, etc.

For example, for the six-color dice, we might assign the following numbers to the various colors:

Color	Assigned value
Blue	3
Red	4
Green	2
Yellow	3
Brown	3
White	2

Note that different colors might or might not have different numbers.

We now define the random number R on each of the possible outcomes as the number we assigned to the outcome w. We write this as: $R(Blue) = 3$, $R(Red) = 4$, $R(Green) = 2$, $R(Yellow) = 3$, $R(Brown) = 3$ and $R(White) = 2$.

Now I ask you what the probability is of finding the color green in this experiment. If the dice is fair, you should say 1/6. However, what if I ask you what the probability is that the outcome is such that the random value defined above is 2? In this case, you look at the table and you will see that there are *two* elementary events for which the R value is 2. These are the yellow and the white outcomes. So your answer should be $2/6 = 1/3$.

We write this answer as:

$$Pr(R = 2) = \frac{1}{3}.$$

This notation reads: The probability that the random variable R attains the value 2 is 1/3.

We can derive a new table from the previous table. In this new table, we have only three entries according to the three possible values of R. These are:

$$Pr(R = 2) = \frac{2}{6} = \frac{1}{3}$$

$$Pr(R = 3) = \frac{3}{6} = \frac{1}{2}$$

$$Pr(R = 4) = \frac{1}{6}.$$

Note also that the sum of the probabilities in this second table must be one.

Now that we have assigned numerical values to each of the outcomes of the experiment, we can define the *average* of the *random variable*. In this case the average is:

$$\frac{1}{3} \times 2 + \frac{1}{2} \times 3 + \frac{1}{6} \times 4 = \frac{17}{6}.$$

In general, the average of the random variable R is simply the sum over all possible *values* of R times the probabilities of occurrence of these values.

6.7 Summary of What We Have Learned in This Session

This session dealt with some concepts that you probably knew intuitively. I believe that you already knew what an average quantity was, and how to define it. I have also introduced two quantities that you might not have heard of before — the standard deviation and random variables. These, as well as many other concepts, are important in the mathematical theory of probability and in its applications. However, these concepts are not essential for the purposes of understanding this elementary book.

Session 7

Probability Distributions

If you are the owner of a jeans company and you export your goods to different countries, it would be wise for you to know something about the *distribution* of heights of men in different countries. Suppose you only know that the *average* height of men in Tall-Land is about 1.8 m, and it is about 1.5 m in Short-Land. It would not be profitable for you to produce only two sizes of pants to fit the men in Tall-Land and Short-Land. You might even lose money, since perhaps no one in those countries has the exact heights of 1.8 m and 1.5 m, respectively (see Fig. 7.1).

It is somewhat better if you also know the *spread* or the standard deviation from the average. However, the best thing to know is the *distribution* of heights in each country to which you export your jeans. Knowing the distribution of heights will help you to produce jeans with a distribution of sizes that will fit the needs of the specific country.

Similarly, the knowledge of the distribution of the number of cars passing an intersection at different hours of the day will help in the programming of the best time intervals for the red and green traffic lights at that intersection.

There are, of course, many other examples for which the knowledge of the distribution of some events might be useful in

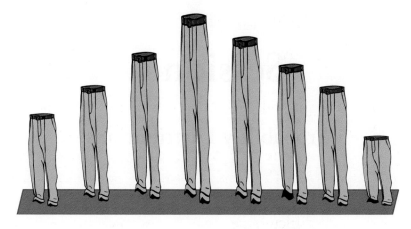

Fig. 7.1   Pants of different sizes.

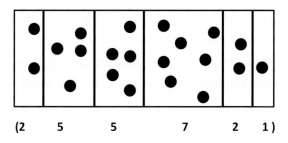

(2   5   5   7   2   1)

Fig. 7.2   Distribution of marbles in different cells.

helping to plan and program for some activities associated with these events. In this session, we shall discuss some theoretical aspects of probability distributions, as well as present some of the most common distributions that we encounter in either our daily lives or in the sciences.

We already know what a distribution is. You have N marbles distributed in k cells, such that there are N_1 marbles in cell 1, N_2 marbles in cell 2, and so on, until there are N_k marbles in cell k. The k numbers $N_1, N_2, \ldots, N_k$ describe the *distribution* of the marbles in the cells (Fig. 7.2). Similarly, we can define the fractions of marbles in each cell by the k numbers:

$$x_1 = \frac{N_1}{N}, \quad x_2 = \frac{N_2}{N}, \ldots, x_k = \frac{N_k}{N}$$

where the sum of x_i (i from 1 to k) must be equal to one. These fractions are also probabilities. Therefore, the set of numbers $x_1, \ldots, x_k$ can be referred to as a *probability distribution*. Can you describe the *event* for which x_6 is the probability of its occurrence?

You should be able to answer this question. But just in case you are having difficulty, let me suggest one possible answer. Suppose the marbles are labeled with different numbers from 1 to N. I pour all of the marbles into a box with k cells. What is the probability of picking while blindfolded a specific *marble* that is labeled, say, "4" from all of the N marbles? This is $Pr(specific\ marble) = \frac{1}{N}$. Note that we assume equal probabilities for picking any of the specific marbles. Next, we ask: What is the probability of picking that specific marble that is labeled "4" from cell number "6"? This time the answer is: $\frac{N_6}{N} = x_6$. The larger the fraction x_6, the larger the probability of finding that specific marble in cell "6." Clearly, if $N_6 = 0$, then this probability is zero, while if $N_6 = N$, then this probability is one.

In general, if we have an experiment with n outcomes: A_1, $A_2, \ldots, A_n$ and the corresponding probabilities $p_1, p_2, \ldots, p_n$, we say that we have the *probability distribution* of these experiments. Note that in the mathematical theory of probability, a distribution is defined slightly differently from how we have defined it. But we shall disregard this difference. There is one more thing to note that will keep you thinking before we discuss some specific distributions. In most cases in daily life, as well as in the sciences, we get a distribution either by collecting statistics (heights of persons, temperatures at different days in cities, etc.)

or from some other arguments (outcomes of a dice or a coin). In some cases in physics, we also talk about the probability of a probability distribution. This sounds awkward but it is not. We shall discuss these kinds of probabilities and their connection to the Second Law of Thermodynamics in next session.

7.1 The Uniform Distribution

Have you even wondered how the fragrance from a small drop of perfume dropped in the corner of a room will, after a short time, permeate to the entire room? Of course, the reason the molecules of the perfume reach the receptors on the inner part of the nose is that they possess kinetic energy. Remember the distribution of velocities from the previous session. Indeed, the kinetic energies of the molecules propel them to move randomly in all kinds of directions and all kinds of velocities. But why do they not stay within the drop of perfume, or in its vicinity? The fact is that after a very short time, the molecules of the perfume will reach a uniform distribution throughout the entire room. This means that the probability of finding any of these molecules in a small region anywhere in the room is the same (disregarding for the moment the effect of gravity on the molecules). If you open a window and a gust of wind blows into the room, at that very moment, the probability of finding any molecule in a given region in the room is not constant. The wind has caused a turbulent motion of the air in the room. Wait a few minutes and the air will settle into a uniform distribution again. You may take this fact for granted, but understanding it is far from trivial. If you have learned about thermodynamics, then you would say that the reason for this uniform distribution is the Second Law of Thermodynamics. This is true; however, underlying the Second Law is a more fundamental principle. The *uniform probability distribution* is the distribution that has the largest *probability* (provided we neglect the gravitational field). I will return to this aspect of the uniform distribution in next session. By uniform distribution, we mean that a molecule can be found in any small element of volume with equal probability. The "largest probability" refers to the *probability* of the probability distribution. This will become clear shortly.

Let us do a simple experiment. Suppose a box is divided into, say, 10 cells. Within one of these cells are placed 10 marbles. The partitions between the cells are high enough so that all 10 of the marbles can be contained in one of the cells. A two-dimensional illustration is shown in Fig. 7.3. Initially, we place all 10 of the marbles in one cell, as shown in Fig. 7.3a.

We next shake the box vigorously in all directions in such a way that the marbles can cross over the partitions between the cells. While we are shaking the box, we take snapshots of the *distribution* of the marbles in the various cells. A distribution of the 10 marbles in the 10 cells is the set of numbers:

$$N_1, N_2, N_3, \ldots, N_{10}$$

where N_1 is the number of marbles in cell "1," N_2 is the number of marbles in cell "2," and so on. We sometimes call the distribution of the marbles in the cell a *configuration*. The

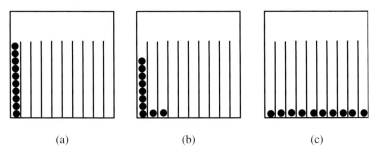

(a)   (b)   (c)

Fig. 7.3   (a) The initial distribution, (b) an intermediate distribution and (c) the distribution after a long time.

initial configuration is written as:

$$N_1 = 10, \quad N_2 = 0, \quad N_3 = 0, \ldots, N_{10} = 0.$$

This means that all of the marbles are in cell "1." When we start shaking the box, we expect that some marbles from cell "1" will cross over into cell "2" or "3," and so forth. For instance, after a few shakings of the box, we might find the configuration:

$$N_1 = 8, \quad N_2 = 1, \quad N_3 = 1, \quad N_4 = 0, \ldots, N_{10} = 0.$$

This means that two of the marbles have crossed over the partitions; one reached cell "2" and one reached cell "3" (Fig. 7.3b).

Imagine that, while shaking the box, you have taken millions of snapshots. Each snapshot shows a specific configuration of the marbles in the cells. Each *configuration* defines a probability distribution $x_1 = \frac{N_1}{N}, x_2 = \frac{N_2}{N}$, and so on. For any *given* configuration, we can ask: What is the probability of finding a *specific* marble, say the one numbered "7," in a *specific* cell, say cell number "4"? At this moment, we are not interested in this probability, but we are interested in a different probability. While shaking the box, we record all of the distributions of marbles in the cells: $N_1, N_2, \ldots, N_{10}$. Each distribution defines a *probability distribution*: $x_1, x_2, \ldots, x_{10}$. Now we ask: What will be the most *probable* distribution of the marbles in the cells? In other words, the probability distributions we recorded are now the "new events." Here are some examples: (The first one is the initial distribution):

$$(10, 0, 0, 0, 0, 0, 0, 0, 0, 0)$$
$$(9, 1, 0, 0, 0, 0, 0, 0, 0, 0)$$
$$(8, 1, 1, 0, 0, 0, 0, 0, 0, 0)$$

$$(10, 0, 0, 0, 0, 0, 0, 0, 0, 0)$$
$$(6, 1, 0, 1, 0, 1, 1, 0, 0, 0)$$
$$(4, 0, 1, 1, 1, 1, 0, 1, 1, 0)$$
$$(2, 0, 1, 1, 1, 1, 1, 1, 1, 1)$$
$$(1, 1, 1, 1, 1, 1, 1, 1, 1, 1)$$
$$(1, 2, 0, 1, 1, 1, 1, 1, 1, 1)$$
$$(1, 1, 1, 1, 2, 0, 1, 1, 1, 1).$$

From these events, we count the fraction of each *distribution*, and we ask: What is the *distribution* that is most likely to occur or have the highest probability?

We shall come back to this question in the next session. Meanwhile, try the following thought experiment.

Suppose we repeat the same experiment as above, but now we start with three blue marbles numbered 1 to 3, five red marbles numbered 1 to 5 and two green marbles numbered 1 and 2. We shake the box vigorously, and take millions of snapshots.

What is the most probable distribution of marbles in the cells?[1]

7.2 The Bernoulli Distribution and the Binomial Distribution

Consider an experiment for which there are *only two* possible outcomes: H or T in tossing a coin, hitting the right (R) or the left (L) area on a board or finding a molecule in one or the other of a pair of compartments.

We assume that the probability of the occurrence of one of the two outcomes is p and the second is q. Since we always assume that the experiment is carried out and one of the outcomes has occurred, we must have $p + q = 1$.

Now we repeat the same experiment 10 times. What is the probability of the *specific* sequence of outcomes

H H T H T T H T H H

given that the probability of H is p and the probability of T is q (we use the language of H and T, but you can use R or L or any other notation for the two events) and given that the sequence of experiments are independent? We have:

$$Pr(a\ specific\ sequence\ of\ six\ Hs\ and\ four\ Ts)$$
$$= p \times p \times p \times p \times p \times p \times q \times q \times q \times q$$
$$= p^6 q^4.$$

We used here the rule that the probability of the 10 independent events is a *product* of all of the individual events.

In general, for N experiments of the same kind as before, the probability of obtaining a *specific sequence* of n outcomes H and $N - n$ outcomes T is:

$$Pr(a\ specific\ sequence\ of\ n\ Hs\ and\ (N - n)\ Ts) = p^n q^{N-n}.$$
$$(7.1)$$

Note that I have emphasized the words "specific sequence" in the sentence above. There is a subtle point to be aware of before we proceed to the next step. A *specific* sequence is when we specify the first outcome, the second outcome, and so on, until the Nth outcome. For example, for four experiments, the *specific* example HTTT has probability:

$$Pr(HTTT) = pq^2.$$

The *specific* sequence THTT also has the same probability:

$$Pr(THTT) = pq^2.$$

Any *specific* sequence having one H and three Ts has the same probability pq^3. When we say *specific* sequence, we mean that we are given the specific order of the events; e.g. first T, second H, third T and fourth T. We see that for *each specific* sequence having one H and three Ts, the probability is pq^3, independently of the specific order. This result follows from the assumption of *independence* of events and from the application of the multiplication rule for the probabilities of independent events.

Although it sounds a little paradoxical, each *specific* sequence having one H and three Ts has the same probability (pq^3) no matter what the specific *order* is. If you are not convinced, write down a few *specific* sequences of four results and calculate the probability for each of these sequences.

Now we ask a slightly different question: What is the probability of *any sequence* of four outcomes, one of which is H and three of which are Ts? The emphasis is now on the words *any sequence*. To calculate this probability, we first write *all* of the possible *specific sequences*. These are:

$$HTTT, \quad THTT, \quad TTHT, \quad TTTH.$$

Check that these are *all* of the possible *specific* sequences with one H and three Ts. The question we ask is about the probability of finding either the first, the second, the third or the fourth of these specific sequences. To calculate this probability, we have to use the rule of *summing* the probabilities of *disjoint* events. The *sequence* can be *either* the first, or the second, or the third or the fourth. We cannot obtain two of these at the same time. Hence, the four listed events are disjoint, and the probability of obtaining any of the events in this list is:

$$Pr(\textit{any sequence of one H and three Ts}) = 4pq^3.$$

Exercise: Calculate the probability of the specific sequence HHHTTT and the specific sequence HTHTHT. What is the probability of a *specific* sequence of six outcomes having three Hs and three Ts? What is the probability of *any* sequence of six outcomes having a three Hs and three Ts?[2]

The generalization for a sequence of N throws with n Hs and $(N - n)$ Ts is straightforward:

$$Pr(any\ sequence\ of\ n\ Hs\ and\ (N - n)\ Ts) = \binom{N}{n} p^n q^{N-n}.$$

(7.2)

Equations (7.1) and (7.2) are very important. You should try to calculate in detail some simple examples, then generalize without looking at Equations (7.1) and (7.2). I suggest you carry out the following two examples. For each of the following experiments, calculate the probability of a *specific* sequence and of *any* sequence:

(1) A sequence of four outcomes with two Hs and two Ts.
(2) A sequence of five outcomes with four Hs and one T.

Let us do some calculations for very large number of marbles, or particles. In the following example, we use the language of particles (atoms or molecules) found in one compartment or in another. This is a very important case in the theory of ideal gases.

We have a box that is divided into two equal-volume compartments, right (R) and left (L), (Fig. 7.4). The probability of finding a single particle in R is $p = 1/2$, and in L the probability is $q = 1/2$. For the present purpose, "ideal gas" means that the particles are independent. The probability of finding a specific particle in R or in L is independent of the number of particles actually being in R or in L. Note that if

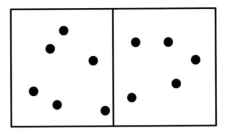

Fig. 7.4 A box divided into two equal volume compartments.

there are interactions between the particles, then this assumption about independence is not valid.

Suppose we have four particles in the box. The probability of finding one specific particle in R and the other three in L is the same as the probability we have found for the sequence HTTT; we have simply replaced the event "H" with the event "R," and the event "T" with the event L. Thus, the probability of this event is:

$$pq^3 = \frac{1}{2} \times \left(\frac{1}{2}\right)^3 = \left(\frac{1}{2}\right)^4 = \frac{1}{16}.$$

In the case of a coin with H and T, or with marbles, we moved from a *specific* sequence to *any* sequence, as we did from Equation (7.1) to Equation (7.2). In the world of particles, we cannot distinguish between the individual particles; the particles cannot be labeled. Therefore, the only question we are interested in is: What is the probability of finding n particles (not specifying which ones) in R and the rest $(N - n)$ in L?

The probability of this event is the same as in Equation (7.2), but for $p = q = 1/2$. We write it again as:

$$Pr(n \text{ in } R \text{ and } (N - n) \text{ in } L) = \binom{N}{n} \left(\frac{1}{2}\right)^n \left(\frac{1}{2}\right)^{N-n}$$

$$= \frac{N!}{n!(N - n)!} \left(\frac{1}{2}\right)^N. \tag{7.3}$$

This is a very important formula. It is based on pure logic. In deriving this formula, we used the *multiplication rule* for N *independent* events to obtain the factor $\left(\frac{1}{2}\right)^{N}$. Then we used the *sum rule* for *disjoint* events; we have exactly $\binom{N}{n}$ events of this kind (n specific particles in R and the rest in L), and all of these are disjoint events. Since each has the same probability $\left(\frac{1}{2}\right)^{N}$, the sum over all of these disjoint events gives $\binom{N}{n}$ times the quantity $\left(\frac{1}{2}\right)^{N}$, which is the result in formula (7.3).

Again, I urge you to check this general formula for some small numbers of N and n.

Figure 7.5 shows the probability $Pr(n \text{ in } R \text{ and } N - n \text{ in } L)$ for different values of N. We use the shorthand notation $Pr(n, N - n)$ for "n in R and $N - n$ in L." For example, when we have four particles, then for $n = 0, 1, 2, 3, 4$, the probabilities are:

$$\frac{1}{16}, \frac{4}{16}, \frac{6}{16}, \frac{4}{16}, \frac{1}{16}$$

respectively.[3]

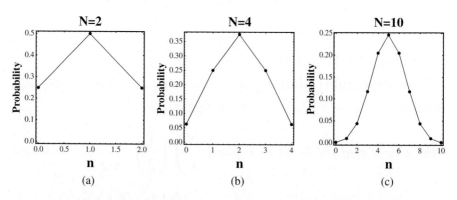

Fig. 7.5 The probabilities $Pr(n, N - n)$ for small values of N.

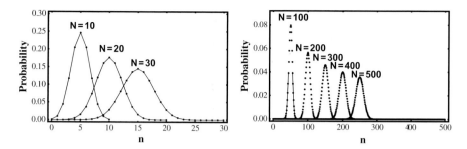

Fig. 7.6 The probabilities $Pr(n, N - n)$ for large values of N.

Figure 7.6 shows again the probabilities $Pr(n, N - n)$ for larger values of N. We see that in each case there is a maximum probability at about $n = N/2$. Note that in all of these cases, we have a finite number of points in the graph. This result is important. For any given N, there are many distributions of the particles between the two compartments. However, there is one distribution for which the probability is maximal. The value of n at which the maximum occurs is always at $n = N/2$ (for even values of N). This is again a very *precise rule* that we obtain from a completely random experiment. You should also note that the maximum of $Pr(n, N - n)$ at $n = N/2$ corresponds to the uniform distribution of particles between the two compartments (think of the two compartments as two cells in the previous experiment with marbles in the cells).

Exercise: Calculate the probability $Pr(n, N - n)$ for $N = 6$ and all possible values of n in a box with two compartments having different sizes, say $p = 1/3$ and $q = 2/3$ (see Note 4).

The occurrence of a sharp maximum of $Pr(n, N - n)$ for large N is important in connection with the Second Law of Thermodynamics (see Session 8). Let us take a slightly different view of this fact. Suppose I start with any initial value of n and $N - n$. If this is a box with marbles in it, then I will shake it vigorously and follow how n changes with time. If this is a system of particles, then I do not need to do any shaking; the particles will do the job of "shaking" by moving about at random. In any case, if we follow how the number n changes with time, we shall always find that, after some time, n reaches a value near $N/2$. You can try this yourself, or just imagine that you are doing the experiment.[5]

Note that when N becomes larger, the shape of the curve become smoother and has a typical bell-shaped curve. We shall see the origin of this particular shape in the next subsection.

The conclusion from this subsection is very important. Starting from any initial distribution of n and $N - n$, the system will evolve with high probability towards the uniform distribution, i.e. $n = N - n \approx N/2$. Here we see the "uniformity" only with respect to the two compartments, but the conclusion is valid for any division of the entire volume into any number of cells.

Now think of a real system — a small glass of water with a volume of about 20 cm^3. In this glass, there is a huge number of water molecules, something of the order of 10^{23} (this means one followed by 23 zeroes. Can you imagine how big this number is?). Imagine that the glass is divided into, say, 1000 cells. Still, each cell will contain a huge number of molecules. In this mental experiment, you do not need to "shake" the molecules. The molecules "shake" themselves. You do not need to actually do the division of the total volume into 1000 cells, just imagine doing it.

In this system, if you record the configurations of the molecules in the cells, you will find that the *most probable configuration* will be the one wherein there are about 10^{20} water molecules in each cell. This result that you have just imagined is one manifestation of the Second Law of Thermodynamics (discussed in Session 8).

Look again at Fig. 7.6 and you will see that the maximal value of the probability $Pr(n, N - n)$ becomes lower when N increases. This means that the probability of obtaining the exact value of $n = N/2$ becomes *smaller* as we take larger and larger value of N. However, we are not interested in the *exact* value $n = N/2$, but for values of n in the *neighborhood* of $N/2$, say ($n = N/2 \pm 0.01N$; see example in Fig. 7.7), the probability becomes almost 1 for large N. This means that when N is large, the probability of finding the system with the exact $n = N/2$ is not large, but when we allow $n = N/2 \pm 0.01N$, the probability is very nearly equal to *one*. This is shown in Fig. 7.8. This fact is important in connection with the Second Law of Thermodynamics, as will be discussed in Session 8.

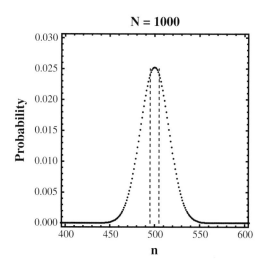

Fig. 7.7 The probabilities $Pr(n, N - n)$ for $N = 1000$. We are interested in values of n "near" the value of $n = N/2$.

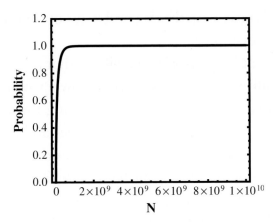

Fig. 7.8 The probability of finding the values of $n = \frac{N}{2} \pm 0.01N$. (See Fig. 7.7 for different values of N).

7.3 The Exponential Distribution

The exponential distribution is one of the most important distributions in physics. In physics, it is usually referred to as the Boltzmann distribution. When discussing the uniform distribution, we emphasized that our system (of marbles or molecules of a gas) is not subjected to an external field, say gravity.

In this section, we examine the effect of gravity or any other external field on the distribution of molecules in a column of a gas of height h.

Suppose we have a system of marbles in a vertical column (Fig. 7.9). We shake the column for a while and measure the average density (i.e. the number of marbles per unit of volume) of marbles at each location in the column. We shall find that, after some time, the density is uniform throughout the column. The density is defined as $\rho = \frac{N}{V}$ (N is the total number of particles and V is the total volume).

However, when the same column is placed vertically in a gravitational field, the resulting distribution of marbles will

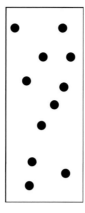

Fig. 7.9 Marbles in a vertical column with no gravitation.

Fig. 7.10 Marbles in a vertical column in a gravitation field.

always be such that they will fall down and accumulate layer upon layer on the floor of the column (Fig. 7.10).

We next perform the same experiment with gas in the column. In this case, we need not shake the molecules as we did with the marbles. The particles "shake themselves" from the inside. In other words, the same kinetic energy that causes the spread of molecules in the gas will be the "shaking agent."

We start with a total of N molecules in a very high column rising vertically from the ground ($h = 0$) to 1 km high ($h = 1$ km). In the absence of an external field, you already know that the average number of molecules in any small box of size v is simply $\frac{Nv}{V}$. If we divide the total volume V into smalls strips of volume v each, the average number of particles in v is ρv where ρ is the density of particles in the entire system $\rho = \frac{N}{V}$.

Next, we "turn on" the gravitational field. What do you expect to happen (we assume that the walls of the box are solid, impermeable to molecules and are perfectly insulated)? Remember that the molecules are still rushing around in all directions and have some distribution of speeds, as we have seen in Session 6 (and we shall soon encounter again in the next subsection). Therefore, there is a tendency as before for each molecule to reach any point in any strip along the column.

However, we intuitively feel that because of the gravitational field, the molecules will be attracted downwards. In other words, the density of molecules at a higher level in the column will be less than the density at a lower level (Fig. 7.11). This fact has been known for a long time. It was used by pilots to measure

Fig. 7.11 Gas in a vertical column in a gravitational field.

their height above ground. By simply measuring the density of air outside an airplane, one can calculate its height from the density. Similarly, many laboratories use centrifuges to separate heavy particles from light ones. In a centrifuge, there is a strong force that pushes the particles away from the center of rotation.

The fact that at higher levels in the column the density of molecules is lower is well known and intuitively clear. What interests us here is the *distribution* of the molecules along the column in a gravitational field.

The answer to this question is far from being trivial. We can easily guess that the molecules will be attracted downwards, but it is far from easy to guess or to calculate the actual "density profile" along the column at equilibrium. To find the actual density profile, one needs some mathematical tools. I will not bother you with this here. In statistical mechanics, one derives the density profile as was done by Boltzmann. The result is:

$$\rho(h) = \rho(0) \exp\left[-\frac{mgh}{k_B T}\right]$$

where $\rho(0)$ is the density at level zero, $\rho(h)$ is the density at level h, m is the mass of the particles, g is the gravitational constant, T is the absolute temperature and k_B is the constant known as the Boltzmann constant. The shape of this function is shown in Fig. 7.12. I should mention here that a more elegant way to derive this distribution is to use the Shannon measure of information, which we shall discuss in next session. There is also an experimental method of deriving this distribution. A simple experiment with marbles in cells is presented in Ben-Naim (2010). In that book, you can also find a qualitative explanation as to why this particular distribution is obtained. We shall not need to go into such details in the present book.

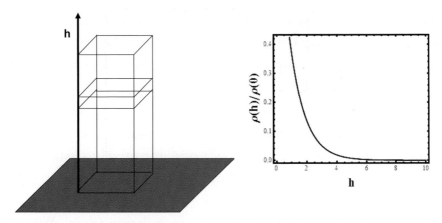

Fig. 7.12 The density of the gas at different heights also known as the Barometric distribution.

An elegant proof of the exponential distribution is available, but it requires some mathematical proficiency. We should also mention here that the eventual exponential (or Boltzmann) distribution is the distribution that is the *most probable* distribution. Again, with mathematics, one can prove this contention.[6]

We have discussed the distribution of molecules in a gravitational field. Another important application of the Boltzmann distribution is in molecular physics. Suppose that molecules can occupy a series of discrete energy levels: $\varepsilon = 0$, $\varepsilon = 1$, $\varepsilon = 2, \ldots$, and so on. Suppose also that there is no limit on the number of molecules occupying any of these levels. The probability of finding any molecule at an energy level ε is proportional to the quantity $\exp\left[-\frac{\varepsilon}{k_B T}\right]$, where again k_B is the Boltzmann constant, T the absolute temperature. Figure 7.13 shows a few curves for different temperatures. The higher the temperature, the more spreading of the molecules is observed on a larger range of energy levels. At very low temperatures, most molecules accumulate at the lowest energy level.

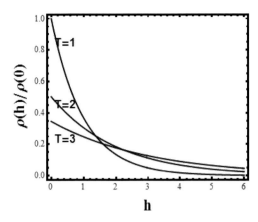

Fig. 7.13 Density distribution in a gravitational field for different temperatures.

7.4 The Normal Distribution

We opened the previous section by claiming that the exponential (or Boltzmann) distribution is one of the most important distributions in physics. This is true, yet there is an even more important distribution that occurs everywhere — not only in physics, but almost anywhere in which statistics applies. This is why it is referred to as the *normal* distribution. It is the *norm* rather than the exception. This distribution is referred to by different names, such as the bell-shaped function (because of its typical shape), the Maxwell–Boltzmann distribution of velocities in one dimension and the Gaussian distribution after Carl Friedrich Gauss (1777–1855).[7]

There are many ways of deriving this particular distribution. All require some degree of mathematics. However, one method of obtaining a *feel* for the shape of this distribution has already been discussed in the section on the binomial distribution. We saw that when N particles are distributed in two compartments. The probability distribution $Pr(n, N - n)$ of finding n particles on the left (or n Hs in N series of coin tosses) and $N - n$ on the right tends to have a bell-shaped form. One can prove mathematically that in the limit of very large N, the binomial distribution will tend to the normal distribution.[8]

A more elegant derivation based on Shannon's measure of information (see Session 8) is available, but this again requires some mathematics.[8]

There is also an "experimental" way of obtaining the normal distribution, which you can see by either imagining an experiment with marbles in cells or simulating the experiment on a computer. This is also described in Ben-Naim (2010).

The normal distribution was first discovered by analyzing the distribution of errors in an experiment. If you measure any quantity, say the heights of persons in a given city, the concentration of sugar in your blood or the weights of newly born babies, you will find a distribution that is similar to the bell-shaped curve. In this distribution, you can see that there is a value that has a maximum probability (this value may or may not coincide with the average value; in a normal distribution, the average and the most probable values are the same). A characteristic feature of the normal distribution is that the probability of the deviations from the average value becomes smaller and smaller as we get farther and farther from the average value.

The simplest way of visualizing the normal distribution is in the distribution of velocities of particles in a *one-dimensional* system. We assume that particles have kinetic energy of motion and that the total energy of all of the particles is constant. We also assume that there are no external fields that will affect the location or the velocities of the particles. Due to random collisions between the particles, we expect that motion to the right is as probable as motion to the left, and whatever

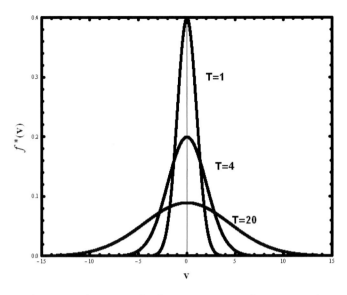

Fig. 7.14 Distribution of velocities in one-dimensional system.

distribution of velocities is, it will be symmetric about the value of $v = 0$ (Fig. 7.14).

The larger the deviation from the center, the lower the probability of finding a molecule, with such a velocity. As we have pointed out earlier, the width of the curve as a measure of the standard deviation is proportional to the temperature. The higher the temperature, the greater the spread of the molecules in a larger range of velocities.

There is one important aspect of the velocity distribution that is different from the locational distribution of molecules in a gas.

In the section on the uniform distribution, we found that in order to obtain the uniform distribution, we needed to *shake* the system of marbles or any kind of particle. In the world of atoms and molecules, this shaking is done from "inside" the system, i.e. due to the random motions of the particles. Thus, the random motions of the particles bring about the uniform distribution of locations.

In the case of velocities, the random motions of the particles play a double role. First, the normal distribution is the eventual equilibrium distribution of velocities. Second, the eventual distribution of velocities is brought about by the random motions of the particles.

Finally, it should be noted that, so far, we have discussed the probability distribution of *velocities* in a one-dimensional system. In this case, we have a symmetrical normal curve; it is as probable that a particle will move in the positive direction as in the negative direction. In Session 6, we discussed the distribution of *speeds* of particles. The speed is the absolute value of the velocity of a particle in any direction in space. This speed is, by definition, always positive and its distribution can be derived from the normal distribution in one dimension. Clearly, in this case, the distribution is not symmetric. It has the form as shown in Fig. 6.5 and it is referred to as the Maxwell–Boltzmann distribution.

7.5 Conclusion

In this section, we have acquainted ourselves with the most important distributions and the most frequently found distributions in natural phenomena. We did not provide any proofs about these distributions. However, the "proofs" are available both experimentally and theoretically. The experimental proof is easy to carry out, whereas the theoretical proof is more difficult to obtain. The latter requires some sophisticated mathematics. In the next Session, we shall point out that all the three distributions, i.e. the uniform, the exponential and the normal distributions, are all related to the Second Law of Thermodynamics.

Session 8

Shannon's Measure of Information

All that you have learned in this book culminates in this last session. The concept we shall learn about in this session does not regularly feature in textbooks on probability. However, it is a concept that is based exclusively on probabilities. It has the form of an average quantity, but it is a very special and unusual average. It is *defined on* any *probability distribution*, yet it also provides a method for *deriving* probability distributions. It was originally defined as a measure of "information," but it turns out to be much more than that — it is also an average *uncertainty* about the outcomes of any given experiment. Nowadays, it encompasses a wide and rich range of applications in all of the sciences and beyond. In short, this is the most interesting and the most beautiful concept I have ever encountered. I hope you will share some of my excitement, admiration and enthusiasm for this beautiful concept. We shall use the symbol S in honor of Shannon and refer to it as the *Shannon measure* of *information*, or SMI for short.

8.1 Definition of the SMI

For any probability distribution (of an experiment with finite numbers of outcomes) $p_1, p_2, \ldots, p_n$, we *define* the quantity:

$$S = -\sum_{i=1}^{n} p_i \log_2 p_i$$

where $\log_2 p_i$ is the logarithm of p_i with a base 2.

You see, as I told you above, this quantity contains probabilities, only probabilities and nothing but probabilities!

Although you should know by now what each symbol in this formula is, I shall repeat: We consider an experiment having n outcomes, indexed 1 to n. The probability of the ith outcome is p_i. We take the logarithm to the base 2 of this probability, and form the *average*, i.e. the sum over all indices i of the quantity $-p_i \log_2 p_i$.

You might hastily conclude that since it contains only probabilities, then this concept must be a purely probabilistic concept. However, this is not the case. The meanings assigned to this quantity is plentiful, and the applications of this concept are far-reaching, even further than probability theory itself.

Two comments need to be made before we continue:

(1) We shall always define the SMI for a finite number of outcomes. There is an extension of this definition for the infinite and the continuous cases. However, there are some mathematical problems with this extension that we will not deal with here.[1]

(2) Shannon used the letter H to define his quantity. I chose S first to remind you of Shannon's outstanding, perhaps even miraculous achievement, and second, to indicate (and only to indicate) that this quantity is also related to one of the

most mysterious quantities that is most revered and most misunderstood in physics: Entropy.

I have deliberately emphasized that I use the letter S only to *indicate* its *relationship* with entropy. It is unfortunate that many authors of articles and books refer to this quantity as "entropy" or as "informational entropy." It is even more unfortunate that Shannon himself called the quantity that he derived "entropy." This has been the source of great confusion, debate, misunderstanding and misinterpretation of both the SMI and entropy. Therefore, from now on, we shall always refer to the quantity S as the SMI defined on the probability distribution $p_1, p_2, \ldots, p_n$. We shall briefly mention its relationship to entropy in the last section of this session. For all that we shall discuss below, whether you have heard the term entropy or not, S should be read as SMI. It is defined in purely probabilistic quantities, and for the moment it bears no relationship with any concept in physics.

We shall discuss some of its meanings below. Let us now examine its application in some simple experiments with which we are already familiar.

(a) *Tossing a fair coin*

In this case, we know that the probability distribution is $Pr(H) = Pr(T) = 1/2$. Hence, we can immediately calculate the SMI for a fair coin as[2]:

$$S(\textit{fair coin}) = -Pr(H) \log_2 Pr(H) - Pr(T) \log_2 Pr(T)$$

$$= -\frac{1}{2} \log_2 \left(\frac{1}{2}\right) - \frac{1}{2} \log_2 \left(\frac{1}{2}\right) = 1.$$

(b) *Throwing a fair dice*

Here, the probability distribution is:

$$Pr(1) = Pr(2) = Pr(3) = Pr(4) = Pr(5) = Pr(6) = \frac{1}{6}.$$

Using the identity[2] $\log_2(1/6) = -\log_2 6$, we have:

$$S(fair\ dice) = -\sum_{i=1}^{6} Pr(i) \log_2 Pr(i)$$

$$= 6 \times \frac{1}{6} \log_2 6 = \log_2 6 \approx 2.58\ldots.$$

(c) *Throwing a fair dice with eight faces*

Here, again we assume that all eight of the outcomes are equally probable, i.e.

$$Pr(1) = Pr(2) = \cdots = Pr(8) = \frac{1}{8}.$$

Therefore:

$$S(fair\ eight\ faced\ dice) = -\sum_{i=1}^{8} Pr(i) \log_2 Pr(i)$$

$$= 8 \times \frac{1}{8} \log_2 8 = \log_2 8 = 3.$$

Check to see that $2^3 = 2 \times 2 \times 2 = 8$.

Exercise: Calculate the SMI of a fair dice with 16 faces, 32 faces and 64 faces. Can you generalize for any fair dice?

8.2 Properties of the SMI

Let us start with the simplest case of an experiment with two outcomes: H and T in tossing a coin or R and L for a marble landing in one of two compartments, etc. We are given the

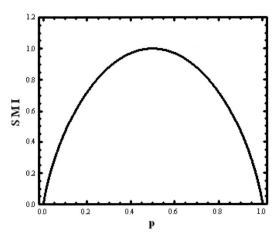

Fig. 8.1   The SMI for the case of an experiment with two outcomes having probabilities p and $1 - p$.

probabilities p_1 and $p_2 = 1 - p_1$, so in effect, we have a one-parameter distribution $(p, 1 - p)$, where $(p = p_1)$. The corresponding SMI for this case is:

$$S = -p_1 \log p_1 - p_2 \log p_2 = -p \log p - (1 - p) \log (1 - p).$$

In this formula, and for the rest of the session, we use log instead of $\log_2$; it will be taken as given that the base is 2.

We see that S is a function of one parameter p. The shape of this function is shown in Fig. 8.1. As can be seen, this function is zero for $p = 0$, and for $p = 1$ it is always positive, and it has a single maximum at $p = 1/2$, at which the value of the SMI is $S = 1$.[3]

We can say that the function S has a maximum at the *uniform distribution* (here $p_1 = p_2 = p = 1/2$) and that it attains the value zero if the distribution is "most non-uniform." By "most non-uniform" I mean that one outcome has probability one, and all the others have probability zero.

We have deduced this property by looking at Fig. 8.1 of S as a function of p for the simplest case. However, this property is shared by the SMI defined for any distribution $p_1, \ldots, p_n$. The SMI always has a *single* maximum at the uniform distribution $p_1 = p_2 = \cdots p_n = 1/n$. It is always a positive quantity and it attains the value of zero at the "most-non-uniform" distribution, i.e. one outcome has probability one and all the others have probability zero.

This property can be easily proven by differential calculus. For the interested reader, an outline of the proof in shown Note 4.

8.3 SMI as a Measure of Uncertainty

Look again at the definition of the SMI. Does it remind you of anything we have already learned? Of course, it must remind you of an *average* quantity. As we have learned, an average of any measurable or observed quantity is defined as follows: Sum over all values of the outcome x times the probability of the outcome x. We write the sum as:

$$\langle x \rangle = \sum_{all\ x} x\ Pr(x).$$

If x is the height of persons, $Pr(x)$ would be the probability of occurrence of that height. If x is the temperature at a specific place at a specific day, $Pr(x)$ would be the probability that the temperature x will occur at that place at a specific day, and so on for any other measurable variable x.

Obviously, the SMI has the *structure* of an average. However, the SMI is a very special average. It is not an average of any *measurable* quantity, but an average of the quantity $\log p$, which is not a measurable quantity. This fact makes the SMI so amazing, mysterious and wonderful. We shall later see how the *entropy* — a quantity defined in thermodynamics that has been a mystery for over 100 years — spawns from the SMI. Therefore, once we comprehend the meaning of the SMI, the mystery associated with the thermodynamic entropy vanishes.

Although the SMI is not an average of any measurable quantity, it is an average nonetheless. But what is it the average of?

Recall that p_i is the probability of observing the outcome i. We interpret it as a measure of, or the uncertainty of, the occurrence of the ith outcome. When $p_i = 1$, we are *certain* that the event i occurred or will occur. When $p_i = 0$, we are *certain* that it will not occur. In between zero and one we can say that the larger the p_i, the more certain we are about the occurrence of the event i.

A caveat

Before we go on, it is important to stress that we are talking about probabilities upon which everyone agrees upon. We used the language of "our uncertainty," or "our ignorance," etc. Indeed, in some usages of probability, the very term "probability" is highly subjective. You may believe that the probability that the messiah will appear tomorrow riding on a white horse is 99%. Others may believe that the probability is not 99%, but rather only 98%, or perhaps 50%, or even 0%. All of these probabilities are highly subjective, but have nothing to do with science. In science, we talk about scientific probabilities upon which we all agree, at least all of those who want to use probability theory. In this sense, the probability 1/6 of a specific outcome of a dice is a measure of the uncertainty of occurrence of that event. It is not considered as a subjective quantity. The uncertainty is not *mine, yours* or *ours*, but rather it is the uncertainty of each one of us who uses the theory of probability. In this sense, it is an objective quantity.

Coming back to SMI: The smaller the p_i, the larger the uncertainty of the event i. Equivalently, the larger $-\log p_i$, the larger the uncertainty about the occurrence of the event i (remember that p_i is a fraction between zero and one; therefore, $\log p_i$ is negative and hence $-\log p_i$ is positive).

Thus, the SMI is the average of $-\log p_i$ or the average *uncertainty* about the occurrence of *all* of the events of an experiment. More briefly, the SMI is the average of the uncertainty about the experiment, or the degree of uncertainty about the outcomes of an experiment.

As a quick verification of this interpretation of SMI, suppose I ask you to guess the outcome of tossing an *unfair* coin. I let you choose between one of the following coins of which we know the distribution of outcomes:

$$\text{Coin A: } Pr(H) = 0.9, \quad Pr(T) = 0.1$$
$$\text{Coin B: } Pr(H) = 0.7, \quad Pr(T) = 0.3$$
$$\text{Coin C: } Pr(H) = 0.5, \quad Pr(T) = 0.5.$$

The rules of the game are very simple. You guess the outcome (either H or T) and we toss the coin you choose (either A, B or C). If the outcome is the same as the one you guessed, you get a dollar; if not, you pay a dollar.

Which coin will you choose and which outcome will you guess, assuming you want to gain and not to lose money?

I can guarantee that if you are older than say 8 years, and if you understood the rules of the game, then you will choose the first coin A and guess the outcome H.

Why am I so sure about what you will choose? After all, you might choose coin A and guess H and *lose*. You might choose coin C, and guess H and *win*. There is nothing that can *guarantee* that you will win if you play only once. All I can say is that there is a *high probability* that if you play with coin A and choose H you will win. Furthermore, if you play the same game with coin A many times, there is a high degree of certainty that you will earn a lot of money. Why? Because the uncertainty of the outcomes of C is larger than that of B, and the uncertainty of B is larger than that of A. It is easy to calculate the SMI for A, B and C — these are 0.47, 0.88 and 1, respectively. This means the uncertainty in the game A is the *lowest*, and it is highest in game C.

The uncertainty interpretation also applies to the more general case when you have *n* outcomes of an experiment. It is sometimes easy to choose a game and sometimes it is not so easy, but the SMI can always give you the measure of uncertainty about the *entire* experiment or the game.

The larger the SMI, the larger the *uncertainty* about the experiment or the game. As we have already seen above, the largest uncertainty is when the probability distribution is *uniform*, and the smallest uncertainty (or the highest certainty) is when the probability is most *non-uniform*, i.e. one event with probability $p_i = 1$, and all others with probability zero. In the last case, you are *certain* about the outcome of the experiment.

8.4 SMI as a Measure of the Amount of Information

The most common interpretation of the SMI is of it being a *measure of information*. This quantity was designed to serve as a measure of information. Before we discuss this interpretation, I should warn you that the SMI has nothing to do with the *information* itself. Information is an extremely general concept.

It can be subjective or objective, it can be important or irrelevant, it can be interesting or dull, etc. All of these attributes of "information" are not relevant to *information theory* in general, and to SMI in particular.

SMI is a measure of the *size* of the information. It is a very special *size* in a very special sense. The following sentence conveys information:

Tomorrow it will rain in Jerusalem.

This information could be important if you plan to take a walk in one of the parks in Jerusalem. To someone in New York, it might have no value at all. Information theory is not concerned with the meaning of this *information*. Instead, it is concerned with some measure of the *size* of the message carrying the information. Obviously, there are different ways of assigning sizes to this message. You can count the number of letters, the number of words or the number of code words you use before you transmit this information in an encrypted way. All these measures are, of course, objective. They do not depend on the meaning, value, importance, content, and so on, of the information.

The measure of information applied to the message above involves the distribution of the letters in the English alphabet. To give you an idea of the corresponding SMI, suppose I want to transmit this sentence by using some "code word" for each letter. One of the simplest codes is Morse code, which assigns a series of dots and dashes to each letter of the alphabet. See Table 8.1 below for details.

Now, suppose that it costs you one cent to transmit a dot and two cents to transmit a dash. Clearly, you would want to assign the shortest code (i.e. the cheapest to transmit) for the most

Table 8.1
The Morse Code

A •−	J •−−−	S •••
B −•••	K −•−	T −
C −•−•	L •−••	U ••−
D −••	M −−	V •••−
E •	N −•	W •−−
F ••−•	O −−−	X −••−
G −−•	P •−−•	Y −•−−
H ••••	Q −−•−	Z −−••
I ••	R •−•	

frequently occurring letter, which in English is "E," and a longer code for the less frequently occurring letters, say "Q" or "Z." In this way, you can minimize your cost for the transmission of any information. The size of the encoded message, or the total cost of the transmission of information, has nothing to do with the *meaning* of the transmitted information. This is one possible *size* of the information. This measure depends on the *probability distribution* of the *letters* of the alphabet in the particular language you are using. The SMI per letter in the English language is defined as:

$$SMI \ (\textit{of letters in the English language}) = -\sum_i p_i \log p_i$$

where p_i is the frequency or the probability of the letter i (see Table 8.2 below and Fig. 8.2) and the sum is over all 26 letters of the English alphabet plus a blank space. As you can see, this quantity has nothing to do with the meaning of the information. It is a purely objective quantity, depending on the characteristics of the language.

Table 8.2 The relative frequencies of letters in English in decreasing order (rounded values).

Letter	Probability (frequency)
Space	0.2
E	0.105
T	0.072
O	0.065
A	0.063
N	0.059
I	0.055
R	0.054
S	0.052
H	0.047
D	0.035
L	0.029
C	0.023
F, U	0.022
M	0.021
P	0.016
Y, W	0.012
G	0.011
B	0.011
V	0.008
K	0.003
X	0.002
J, Q, Z	0.001

8.5 The SMI of a Uniform Distribution

Let us discuss a simpler example in which the meaning of the SMI (not the meaning of the information) should become clear and easy to grasp.

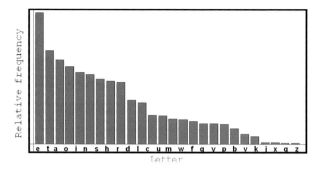

Fig. 8.2 The relative frequencies of letters in English in decreasing order.

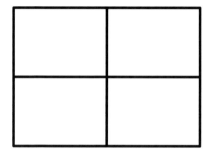

Fig. 8.3 A board divided into four regions having equal areas.

Suppose that I show you a board with four regions having equal areas, as in Fig. 8.3. I throw a dart and I tell you that it hit one of the squares on the board. Your task is to find the square that the dart hit by asking *binary questions*, i.e. I can answer with either "yes" or "no." To make the game more dramatic and more exciting, and perhaps also to encourage you to think harder, let us assume that you pay a dollar for each answer you get. When you correctly determine the square that the dart hit, you get $4.

While you are planning your possible strategy for asking questions, let me tell you that a considerable amount of research has been carried out with this and similar games on children aged 8 to 12. It has been found that younger children tended to ask questions about a *specific* square: "Is it on the upper right square?" etc. Clearly, the younger children were impatient to receive the prize. Of course, they had a chance to win with the first question, paying \$1 and receiving \$4. However, winning in one question has a probability of 1/4 (in this particular game).

On the other hand, older children chose a different strategy, asking, for example, "Is it in one of the two squares on the right-hand side?" Obviously, those who played the game adopting this strategy of questioning could not win on the first question.

However, whatever answer they receive to their first question, they will know the correct location that the dart hit with the second question. Thus, adopting this strategy *guarantees* that with two questions the required information will be obtained (on where the dart hit). Hence, you pay \$2 for the two questions and receive \$4 for finding the location of the dart. If you play this game many times, you could easily get rich.

I will not describe here any more details of this kind of research and the implications regarding the ability of children to choose the better strategy. You can read more about this research in Ben-Naim (2010). We shall refer to the strategy of asking *specific* questions (e.g. "Is it in the first box?") as the "dumbest" strategy, and the one based on dividing the board into two halves (e.g. "Is it in the right half?") as the "smartest" strategy.

If you play this game many times, you will find that, on average, you win more by adopting the second (smartest) strategy than the first. In this particular game, the difference between the two strategies is not too dramatic. But when we increase the number of areas, say from 4 to 8 to 16 and so on, you will find that, on average, the number of questions you will have to ask using the "dumbest strategy" increases with the number of squares, roughly as $N/2$, i.e. *proportional* to N.

On the other hand, adopting the "smartest" strategy of asking questions, i.e. dividing the entire N squares into two halves each

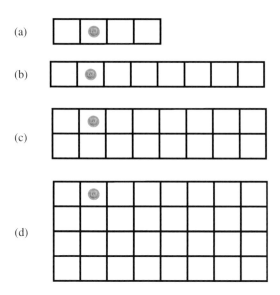

Fig. 8.4 A coin hidden in one of N boxes, with $N = 4$ to $N = 32$.

time, then again into two halves until you find the dart, the number of questions you need to ask is approximately $\log_2 N$, which is much smaller than $N/2$. You can verify this "law" by checking the number of questions you need to ask using the "smartest" strategy for the following cases (Fig. 8.4):

$$N = 4 \quad N = 8 \quad N = 16 \quad N = 32 \quad N = 64.$$

The corresponding number of questions are 2, 3, 4, 5 and 6, respectively.[5]

Note that in this series of games, when we *multiply* the number of squares by two, the number of questions you need to ask increases only by one! In general, for $N = 2^n$ (n integers), the number of questions will be $\log_2 N = \log_2 2^n = n$ (provided you play the game "smartly").

Let us summarize what we have found so far. You are told that a dart has hit one of the N squares. You also know that the areas of the small squares are equal. Therefore, the probability that the dart hit any specific area is $1/N$. You do not know the *information* of where the dart it. So you ask questions in order to get this *information*. We are not really interested in the *information* itself, but rather on how to get it by spending as few dollars as possible, i.e. earning the maximum number of dollars in each game. We have found that the number of questions you need to ask if you are smart enough to adopt the "smartest" strategy is of the order of $\log_2 N$.

As an exercise, suppose that you pay one dollar for each answer you get, and you receive $N/2$ dollars when you find the information (on where the dart hit). How much do you expect to gain on average if you play this game 1000 times, either adopting the "dumbest" strategy (i.e. asking about *specific* squares) or the "smartest" strategy (i.e. dividing into two equal halves each time)? The games use the following values of N:

$$N = 8, 16, 32, 64, 2^{10}, 2^{100}, 2^{1000}.$$

Up to this point, we have found the relationship between the number N of equally probable events (the squares in which the dart hit) and the number of questions for N of the form $N = 2^n$, where n is an integer. We have found that:

number of questions $= \log_2 N$.

One can show that this relationship is valid for any N. We shall not discuss the proof here.[6]

Now that we know the relationship between the number of events N and the number of questions one must ask to find which of the N events has occurred (e.g. where the dart hit), we can ask what this has got to do with *information*.

The information that we do not have is on which event occurred. The number of questions we must ask is a measure of the *size* of that *information*. The larger the number of events, the larger the number of questions we must ask in order to obtain this information. Hence, we use this number as a measure of the *size* of the *missing information*.

8.6 The SMI of a Non-Uniform Distribution

So far, we have discussed events with equal probability or experiments with a *uniform* probability distribution. The next step is to study cases of non-uniform distribution of events. Suppose we have an experiment with N outcomes, each with a different probability. This defines a probability distribution $p_1, p_2, \ldots, p_N$ with $\sum_{i=1}^{N} p_i = 1$. For this distribution, we define the SMI as we did above:

$$S = -\sum_{i=1}^{N} p_i \log p_i.$$

One can prove mathematically that for any given distribution $p_1, \ldots, p_N$, the quantity S defined above is equal (up to an accuracy of ± 1) to the average number of (smart) questions we have to ask in order to find out which event has occurred by asking binary questions.[7]

Thus, the larger the number of questions we need to ask, the larger the size of the information contained in the game. Again, we stress that the information itself is not important. It can be the location of a dart on a board, the location of a coin in N boxes or any other experiment having N outcomes with a given probability distribution $p_1, \ldots, p_N$, in which case the information is about the specific outcome that occurs.

Note that in calculating the SMI of the experiment or of the game, we only used the *given distribution* and not any details of the experiment or the game. Therefore, it is more appropriate to refer to the SMI as a measure of the information *contained in* or *belonging to* a given *distribution*, leaving the details of the experiment unspecified. The distribution can pertain to dice, coins or cards. It does not matter. The only thing that matters is the probability distribution itself. This is why the SMI is so general, and such an exciting quantity.

We shall examine only a few examples here. It is convenient to use again the dart hitting a board divided by N squares, but now the areas are unequal. The probability of hitting the ith region in the board having area A_i is assumed to be equal to $p_i = \frac{A_i}{A}$. The information we need is "where the dart hit" or, for a more general experiment, "the nature of which event occurred." The size of this information is measured by the minimal number of questions we need to ask in order to obtain this information.

Sometimes, the SMI is referred to as a "measure of information" and sometimes as a "measure of the missing information." This distinction is not important. In the first, we assign the "information" to the *experiment*, in the second, we assign the "missing information" to the one who has to ask the questions. Either way, this quantity is an objective quantity associated with a given experiment with a given probability distribution.

Consider the following two divisions of the same board (Fig. 8.5). Both boards are divided into eight regions. On the left (Fig. 8.5a), we have eight *equal* areas, whereas on the right (Fig. 8.5b), we have eight *unequal* areas. Which of these games is easier to play? Putting it differently, suppose you have to find where the dart hit by asking binary questions. Assume that you pay 1 cent for each question you ask, and you will get 10 cents

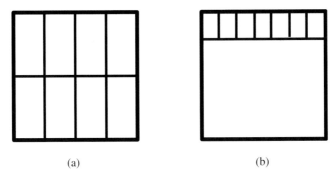

(a)   (b)

Fig. 8.5   A board divided into eight regions.

when you find the region in which the dart hit. The question is now, given the two distributions, in which game can we get the information on "where the dart hit" with the fewest number of questions? The exact answer to this question is *contained* in the values of S for these two cases. The larger S, the larger the *size* of the *information*.

Intuitively, we feel that in this particular example, the game on the right requires fewer numbers of questions. The reason is that with the game on the left we must ask at least three questions. On the other hand, with the game on the right we can ask fewer questions on average. We start by asking, "Is it on the lower area?" There is probability of 3/4 of obtaining a "yes" answer on this first question. If we get a "no" response, we have to ask three more questions on average.

The value of S for these two cases are:

$$S(Left) = \sum_{i=1}^{8} \frac{1}{8} \log 8 = 3$$

$$S(Right) = -\frac{3}{4} \log \frac{3}{4} - \sum_{i=1}^{7} \frac{1}{28} \log \frac{1}{28} = 1.513$$

We see that the S value for the left is larger than for the right. In general, one can prove that the more asymmetric the distribution is, the smaller the size of the information. The maximal value of S is when the distribution is *uniform*. We have already seen that this corresponds to maximal uncertainty about the experiment. Here, we re-interpret maximal uncertainty in terms of the maximum information contained in the distribution of the outcomes of the experiment.

8.7 How Did Shannon Derive the SMI?

Shannon worked at Bell Telephone Laboratories. He was interested in the theory of communication, and more specifically on how to transmit information most efficiently through communication lines.

In 1948, Shannon published a landmark paper entitled "A mathematical theory of communication." In Section 6 of this paper, Shannon writes:

> Suppose we have a set of possible events whose probabilities of occurrence are $p_1, p_2, \ldots, p_n$. These probabilities are known but that is all we know concerning which event will occur. Can we find a measure of how much "choice" is involved in the selection of the event, or how uncertain we are of the outcome? If there is such a measure, say $H(p_1, p_2 \ldots p_n)$, it is reasonable to require of it the following properties:
>
> (1) H should be continuous in the p_i.
> (2) If all the p_i are equal, $p_i = \frac{1}{n}$, then H should be a monotonically increasing function of n. With equally likely events there is more choice, or uncertainty, when there are more possible events.
> (3) If a choice be broken down into two successive choices, the original H should be the weighted sum of the individual values of H.

Then Shannon proved that the only H satisfying the three assumptions above is of the form:

$$H = -K \sum p_i \log p_i.$$

Shannon referred to the quantity he found as the amount of "choice" or "uncertainty" about the outcome. This quantity later became the central concept in information theory and was sometimes referred to simply as "information." Of course, this has caused a considerable amount of confusion, mainly because "information" in general can have *meaning, value, importance,* and so on, but the SMI is a purely objective quantity depending only on the probability distribution, and not on the specific experiment or game.

Over the years, the SMI was re-derived in different ways based on slightly different assumptions. However, it is interesting to note that it was originally derived at a time when it was not even known that such a measure of information existed.

To highlight the gist of Shannon's achievement, let us go back to the 20-question game. I hid a coin in one of N boxes and I tell you that I chose the box where I placed the coin with probability distribution $p_1, \ldots, p_N$. If you prefer, think of the game with the dart hitting a board with N regions, each with relative areas $p_1, \ldots, p_N$. Suppose that you pay a dollar for each answer that you get, and when you find the outcome of the experiment (where the coin is hidden or where the dart hit), you win a prize of X dollars.

How should one play this game using a minimal number of questions so that one maximizes one's earnings? Offhand, it is not clear that such a "maximizing earning method" exists, and even if it does, it is not clear how to determine the number of questions that will give us the maximum returns in the game.

Shannon formulated an equivalent problem, having no idea of whether a solution to his problem existed. He further assumed that if such a measure existed, it must fulfill some plausible properties. With these set of plausible properties, he proved that there is only one quantity that fulfills these properties, and that was how he found the SMI. The details of the proof are highly mathematical and will not concern us here. I simply wanted to convey to you the flavor of the type of problem Shannon faced — to find a solution to a problem without knowing if such a solution existed.

Once the SMI was found, people realized that it could be used in many other fields of research unrelated to communication theory. It was found to be useful in physics and mathematics, biology and psychology, sociology and even in literature, music and other arts. This huge scope of applicability is probably the reason as to why this quantity is so powerful, and why, for some, it is even seen as an awesome quantity.

In this book, we are interested in the theory of probability. I brought the SMI into this book as a quantity that is purely probabilistic, depending only on the given probability distribution, but independent of the experiment or the game that provided this distribution.

Notwithstanding Shannon's enormous achievement, he committed a small semantic mistake — he called his quantity "entropy." Of course, one can call the SMI by any term one wants. Unfortunately, the specific choice of the term "entropy," which was already used in physics, has caused a great amount of confusion, and a vigorous debate ensues as to the very meaning of the SMI, as well as the meaning of entropy.

I shall devote the last section of this book to clarifying the relationship between the SMI and entropy as used in physics. This will be very brief. The reader who is interested in more details is referred to the recommended literature.

Before we discuss the relevance of the SMI to thermodynamics, I owe you an answer regarding the amount of information involved in the three prisoners' problem discussed in Session 5.

As we have noted in Session 5, the "information" used in the three prisoners' problem is not the kind of information used in information theory. However, one can define Shannon's information measure for this problem as follows: Initially, we have three equally likely events, i.e. each prisoner had probability 1/3 of being freed; hence the corresponding SMI is:

$$SMI = -\sum_{i=1}^{3} p_i \log_2 p_i = \log_2 3 \cong 1.585.$$

When prisoner A asks the warden his question and receives his answer, there are only two possibilities left; the corresponding SMI is:

$$SMI = -\frac{1}{3} \log_2 \frac{1}{3} - \frac{2}{3} \log_2 \frac{2}{3} \cong 0.918.$$

In the more general problem with 100 prisoners, we start with a uniform probability distribution, i.e. each prisoner had probability of 1/100 of being freed; hence:

$$SMI = -\sum_{i=1}^{100} p_i \log_2 p_i = \log_2 100 \cong 6.644$$

and after prisoner "1" receives information from the warden, the SMI reduces to:

$$SMI = -\frac{1}{100} \log_2 \frac{1}{100} - \frac{99}{100} \log_2 \frac{99}{100} \cong 0.080.$$

Note that the reduction in the SMI is much larger in the more general problem. *The reduction in the SMI is larger, the larger the number of prisoners.*

Exercise: Solve the problem with 1000 prisoners and calculate the SMI before and after prisoner "1" receives the information from the warden.

Here is one final suggestion as a teasing thought: Suppose that there are 10 prisoners as in the case of Fig. 5.14. Again, prisoner "1" asks the warden the same question as before. The warden does not point to all eight of the prisoners that are to be executed, but only at k prisoners to be executed ($k = 1, 2, \ldots, 8$). The question is again the probability of "1" surviving relative to the probability of survival of one the remaining $9 - k$ prisoners, on which no information is available. Clearly, the larger the value of k, the more "information" is given to prisoner "1." A more detailed discussion of this problem may be found in Ben-Naim (2008).

8.8 SMI, Entropy and the Second Law of Thermodynamics

This section is a culmination of the last session of this book, and the last session is the culmination of the entire book. However, you can skip this section without missing out on anything you could learn on probability theory. But if you are curious about one of the most mysterious quantities in physics, a quantity that resisted understanding for over 100 years and has caused a vigorous debate on its meaning and interpretation, you can stay with me and see how this quantity — entropy — which was defined originally in terms of heat engines and transferred thermal energy, can emerge from SMI.

Consider an ideal gas consisting of N atoms of, say, argon in a volume V and having total energy E. Given the three quantities N, V and E, we say that the system is characterized in macroscopic or thermodynamic terms.

On a molecular level, i.e. in the microscopic world, the description of the system is different. In classical mechanics, we can describe the system by giving the location and the velocity of each atom at each time. Normally, N is of the order of 10^{23}, a huge number of particles. For each atom, we need to specify its location in space by its coordinates x, y and z, as well as the three velocities along these coordinates: v_x, v_y and v_z. Altogether, we need to specify $6N$ parameters at each point in time.

In classical mechanics, there are infinite possible "configurations" for the N particles. This is because each of the $6N$ variables can have an infinite number of values. However, as we pointed out earlier, we are never interested in the *exact point* or the *exact velocity* of each atom. We always restrict ourselves to some small finite "cells" of locations and velocities. For instance, one atom moving along the x axis can have an infinite number of points within the "box," say between 0 to L (Fig. 8.6).

0   L

Fig. 8.6 A one-dimensional box of length L.

In practice, we divide the range $(0, L)$ into a finite number of small cells. The cells can be as small as we wish, but their size is finite, not zero. We are then interested in the cell in which the atom is located, and not the exact point in which the atom is located within the cell. Thus, from an infinite number of points, we obtain a finite number of cells in which the atom can be located.

In what we have said above, we have *voluntarily* divided the entire range $(0, L)$ into a finite number of cells, either because of the limitation of our measuring instruments or simply because we do not care to know the exact location of the atom.

In quantum mechanics, this inaccuracy in the location and velocity of each atom is *imposed* on us. This is called the *uncertainty principle* of quantum mechanics. This principle states that one cannot describe both the location x and the velocity v_x exactly, but within a little "cell" of a size we denote by h, called the *Planck constant*.

We now have a microscopic description of the location and velocity of each particle within a *finite* number of cells. For each atom moving in a three-dimensional space, we have three cells describing the three pairs: (x, v_x), (y, v_y) and (z, v_z). Furthermore, for N of the order of 10^{23}, which is an extremely large number, we have even a larger number of configurations.[8]

Now, I offer you the chance to play the 20-question game in this enormous space. Suppose I take a snapshot of all of the N atoms in the system and record the configuration of all of the atoms, i.e. in which cell each atom is located (remember that each cell contains the location and the velocity of one atom along one axis). I know the specific microscopic description ("configuration" or "state") of the system and you have to find out which state it is in by asking binary questions. How would you play this game?

Before even starting to play you should ask what the probability distribution of the states of the system is. The answer to this question is not simple. In fact, there are many possible distributions. However, there is one specific probability distribution that is called the *equilibrium* distribution. This distribution can be determined either experimentally or mathematically by maximizing the SMI over all possible distributions of the states of the system.

I realize that the last sentence might be difficult to comprehend, so I will say it differently. You are given a 20-question game. In order to play it efficiently (i.e. maximize your earnings), you want to know the distribution of the outcomes. Once you know the distribution, you will be able to calculate the relevant SMI.

One of the outstanding properties of the SMI is that it provides a method of calculating the *most probable* distribution of the states of the system. This is also called the equilibrium distribution of the states.

Once you have the equilibrium distribution of states, you can calculate the SMI pertaining to that particular distribution. Here awaiting you is an outstanding surprise: The SMI that is calculated for the equilibrium distribution of states (which itself was calculated by maximizing the SMI over all possible distributions) is, up to a multiplicative constant, equal to the *entropy* of the system of N atoms in volume V and having total energy E. The multiplicative constant k is the Boltzmann constant (denoted k_B) and we have to change the base of the logarithm. These change only the *scale* of the SMI, not its meaning.

This is a truly astonishing finding. We started with the SMI, which is definable for any probability distribution. The experiment or the game can be tossing a coin or throwing a dice or throwing a dart at a board. For each of these experiments, we can define the corresponding SMI. However, for one particular distribution, the SMI turns into the entropy of the system, the very same entropy that has been defined within the context of heat engines and thermal energy transfer. Astonishing as it may sound, this relationship between the SMI on the one hand and the entropy on the other holds another surprise that is no less astonishing; it unveils the mystery that has enshrouded entropy for over 100 years. "Entropy always increases" — this is called the Second Law of Thermodynamics.

The Second Law of Thermodynamics states that in any spontaneous process in an isolated system (fixed N, fixed volume V and fixed energy E), the entropy always increases. This law has been both a mystery and a subject of vigorous debate. It seems as if there is only a one-way direction in which a spontaneous process in an isolated system can occur. A gas always expands from a small volume to occupy the entire volume V. Heat always flows from a hot to a cold body, and two gases in two separate compartments always mix when a barrier between the compartments is removed (Fig. 8.7).

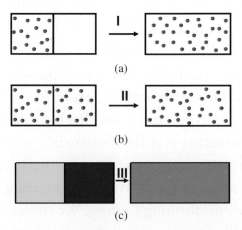

Fig. 8.7 Three spontaneous processes (a) expansion, (b) mixing, and (c) heat transfer.

$$PR(\{p_1,...,p_n\}) = \left(\frac{1}{n}\right)^N \frac{2^{N \times S(\{p_1,...,p_n\})}}{\sqrt{2\pi N^{(n-1)} \prod_{i=1}^{n} p_i}}$$

WHY=2^{N×WHAT}

Fig. 8.8 The relationship between the entropy and the probability, and the symbolic relationship between the questions of "What entropy is," and "Why it always increases."

All of these processes, and many more, we experience in our everyday lives; all proceed in one direction, a direction that has been described as the *arrow of time*. This one-way law seems to be in conflict with the laws of mechanics, which are indifferent to the direction of time.

The debate about this apparent conflict continues to this day. Yet from the point of view of information theory, there is a simple relationship between the SMI of any game and the probability of finding a particular probability distribution. For a system of many atoms and molecules, this relationship connects the entropy of the system and the probability of obtaining the *equilibrium* probability distribution of the microscopic states of the system. This relationship is shown in the formula in Fig. 8.8 for any distribution. It is also shown symbolically as the SMI provides an *interpretation* of entropy; the probability (*Pr*) defined on the probability distribution provides an interpretation of the Second Law, i.e. an answer to the question as to why the system changes in only one direction. This relationship "kills two birds with one stone" — it removes the mystery from both entropy and from the Second Law of Thermodynamics.

Notes

Preface

Note 1: I will use the terms "likelihood" and "probability" interchangeably, as they are used in daily life.

Note 2: The first is the probability that there will be precipitation tomorrow. The second has nothing to do with probability. Relative humidity is a measure of the quantity of moisture in the air, relative to the maximum capacity of the air to hold moisture at any given temperature. Note that the same relative humidity in the summer, say 50%, feels very different from 50% humidity in the winter.

Note 3: List of mathematical notations and terms.

Exponent: For any integer y, we define x^y as the multiplication of x by itself y times. Thus, $5^3 = 5 \times 5 \times 5$. We also note that $x^1 = x$ (e.g. $5^1 = 5$) and $x^0 = 1$. The exponent is also defined for any x and y, but we shall need it only for y being an integer. Also note that $x^{-4} = \frac{1}{x^4}$.

Logarithm: $\log_y x$ is the number you have to insert into the brackets in $y^{(-)}$ so that $y^{(-)} = x$. For example, $\log_2^8$ is 3, since $2^3 = 8$.

Sum: The symbol $\sum_{i=1}^{n} a_i$ means take the sum over all the indices i from 1 to n, i.e.:

$$\sum_{i=1}^{n} a_i = a_1 + a_2 + \cdots + a_n$$

Union and intersection:

$\cup$ reads as *union*, and $A \cup B$ means A *or* B (all the elements belonging to either A *or* B).

$\cap$ reads as *intersection*, and $A \cap B$ means A *and* B (all the elements belonging to both A *and* B).

Factorial: For n integers, $n! = 1 \times 2 \times 3 \times \cdots \times n$. We also define $0! = 1$.

Pr: Probability.

Binomial coefficient: $\binom{N}{n} = \frac{N!}{n!(N-n)!}$.

Session 1

Note 1: The main message of this introductory section to the first session is that there is no bona-fide *definition* of the concept of probability. Some authors admit this from the outset. Others provide methods of *calculating* probabilities, but do not provide a definition of probability.

Note 2: On page 1 of Bennett's (1998) book *Randomness*, you find a quotation from Persi Diaconis (1989): "Our brains are just not wired to do probability problems very well."

I partially agree and partially disagree with this statement. I bring it up here as a first example of a statement that requires one to pause and think: Do you or do you not agree? Do not take words of authority for granted. I personally believe that since we live in a world full of uncertainties, our brains do contain some kind of a rudimentary "sense of probability." I hope that Session 1 convinces you of this. In this sense, I do not agree with the first part of the above cited quotation. However, I agree with the last part of the quotation, i.e. that our brains are not wired to do *difficult* probability problems "very well."

In the same vein, I believe that we have a sense of numbers, or the sense of counting. Of course, this rudimentary sense is not enough to do complicated problems. We have to train our brains to do arithmetic problems "very well."

Note 3: The first statement (i) is a subjective assessment of the likelihood of success or failure. Different persons will reach different conclusions regarding the chances of success or failure. The second statement (ii) would be considered as an objective assignment of probability to the event: outcome {"4"}.

Note 4: See some of the publications by Falk in the bibliography.

Note 5: Most children who participated in this game chose the urn on the right. I am sure you did the same. However, when the children were asked why they had chosen this particular urn, they answered: "Because there are *more blue* marbles in this urn than in the left one."

Note 6: Falk *et al.* (1980 and 2012) concluded that children switched rather early from considering one number rather than two. If we denote by $N(w)$ the number of "winning" marbles, and by $N(l)$ the number of "losing" marbles in the urn, then younger children chose the urn with the larger value of $N(w)$. Some chose the urn having the smaller value of $N(l)$ (these choices were referred to as "one-dimensional" by Falk *et al.*). Older children sometimes made choices by combining the two numbers, either the difference $N(w) - N(l)$, or the ratio $N(w)/N(l)$.

The correct choice is the one made according to which urn has the larger ratio $N(w)/N(l)$ of marbles. This is equivalent to choosing the urn with the larger probability of winning.

$$Pr(win) = \frac{N(w)}{N(w) + N(l)}$$
$$Pr(lose) = \frac{N(l)}{N(w) + N(l)}.$$

Clearly, the larger the ratio $N(w)/N(l)$, the larger the probability of winning.

Note 7: In game A, we assumed that the dice is "fair," which means that all outcomes are equally likely to occur. There is no preferred outcome. Therefore, you cannot "explain" why you have chosen any particular number between 1 and 6. In this game, whatever outcome you choose, say 4 or 6, you can expect to "win," on average, one in six throws. If you play 1000 games, you can expect to "earn" $1000 \times \frac{1}{6} \approx 167$ dollars, independently of the number you have chosen.

In game B, again your earnings are independent of the choice of a particular number. On average, you will earn $1000 \times \frac{1}{6} \approx 167$ dollars if the outcome coincides with the number you chose, but you will lose $1000 \times \frac{5}{6} \times \frac{21}{100} = 175$ dollars. In this case, you will be losing, on average, about $175 - 167 = 8$ dollars after 1000 games.

Note 8: The probabilities and expected earnings are:

For "blue":

$$Pr = \frac{4}{19}, \quad \text{with expected earnings of } \frac{4}{19} \times 2 = \frac{8}{19}.$$

For "red":

$$Pr = \frac{5}{19}, \quad \text{with expected earnings of } \frac{5}{19} \times 3 = \frac{15}{19}.$$

For "green":

$$Pr = \frac{10}{19}, \quad \text{with expected earnings of } \frac{10}{19} \times 1 = \frac{10}{19}.$$

Thus, although the probability of "red" has decreased (since you removed one "red" from the urn), your expected earnings are still larger when you choose red again.

Note 9: The probabilities and expected earnings are:

For "blue":

$$Pr = \frac{4}{18}, \quad \text{with expected earnings of } \frac{4}{18} \times 2 = \frac{8}{18}.$$

For "red":

$$Pr = \frac{4}{18}, \quad \text{with expected earnings of } \frac{4}{18} \times 3 = \frac{12}{18}.$$

For "green":

$$Pr = \frac{10}{18}, \quad \text{with expected earnings of } \frac{10}{18} \times 1 = \frac{10}{18}.$$

Note 10: Statements (a) and (b) are meaningless. There is no probability assigned to an object (see also Session 2). Statements (c) and (d) mean the extent of your belief in certain things provided you know what "Kukuriku" is, which book you are referring to, and what you mean by "interesting." Thus, these statements could be anything from vague to meaningless.

Statements (e) and (f) are meaningful. These still measure the extent of your belief in certain events. However, if you have collected a large amount of data on the weather in certain areas and in certain seasons of the year, or if you have good statistics on the longevity of people in some country and in some period of time, then you might get a good estimate of the probabilities of these events.

The last two statements, (g) and (h), should be accepted by everyone who uses the concept of probability. Basically, these probabilities also measure the extent of our belief in the occurrence of these events. However, these numbers, unlike all the others in statements (a) to (f), can be "verified" experimentally. The word "verified" is used here in a statistical sense or, if you want, you can verify with high "probability" that these probabilities are correct. We shall further discuss the meaning of the previous sentence in next session.

Note 11: Exercise (a) is quite simple. It only requires your sense of probability.

Exercise (b) requires more thinking and more planning. I will let you think about these problems until we get to the end of the last session of this book, where we shall analyze the details of these problems.

Note 12:

(ו) וַיִּבֶז בְּעֵינָיו לִשְׁלֹחַ יָד בְּמָרְדֳּכַי לְבַדּוֹ כִּי הִגִּידוּ לוֹ אֶת עַם מָרְדֳּכָי וַיְבַקֵּשׁ הָמָן לְהַשְׁמִיד אֶת כָּל הַיְּהוּדִים אֲשֶׁר בְּכָל מַלְכוּת אֲחַשְׁוֵרוֹשׁ עַם מָרְדֳּכָי: (ז) בַּחֹדֶשׁ הָרִאשׁוֹן הוּא חֹדֶשׁ נִיסָן בִּשְׁנַת שְׁתֵּים עֶשְׂרֵה לַמֶּלֶךְ אֲחַשְׁוֵרוֹשׁ הִפִּיל פּוּר הוּא הַגּוֹרָל לִפְנֵי הָמָן מִיּוֹם לְיוֹם וּמֵחֹדֶשׁ לְחֹדֶשׁ שְׁנֵים עָשָׂר הוּא חֹדֶשׁ אֲדָר: **מגילת אסתר (ג, ו-ז)**

עַל כֵּן קָרְאוּ לַיָּמִים הָאֵלֶּה פוּרִים עַל שֵׁם הַפּוּר

מגילת אסתר [ג, כו]:

3:6 And he thought scorn to lay hands on Mordecai alone; for they had showed him the people of Mordecai: wherefore Haman sought to destroy all the Jews that were throughout the whole kingdom of Ahasuerus, even the people of Mordecai.

3:7 In the first month, that is, the month Nisan, in the twelfth year of king Ahasuerus, they cast Pur, that is, the lot, before Haman from day to day, and from month to month, to the twelfth month, that is, the month Adar.

Megillat Esther (Book of Esther, 3:6,3:7))

There are other stories in which people cast lots in order to make difficult decisions. Believing that the outcome of the lot is controlled by a divine power, this "mechanism" of decision-making freed people who cast lots from the responsibility of the consequences of their decisions.

Session 2

Note 1: The answers are:

$$\text{(a) } \frac{1}{6}, \quad \text{(b) } \frac{3}{6} = \frac{1}{2}, \quad \text{(c) } \frac{2}{6} = \frac{1}{3}.$$

Note 2: The reasoning here is similar to that which we carried out in calculating the probability of the event "even." In this case, we divide all of the possible events into three groups; $\{1, 2\}$ or "red," $\{3, 4\}$ or "blue," and $\{5, 6\}$ or "green." The required event $\{5, 6\}$ is the same as the outcome "green." Since there are three equally likely colors — red, blue and green — the probability of the occurrence of the color "green" is 1/3.

Note 3: See Fischbein and Schnarch (1997).

Note 4:

$$\text{Number of } \{\text{red, red}\} = NR \times \frac{NR - 1}{2}$$

$$\text{Number of } \{\text{blue, blue}\} = NB \times \frac{NB - 1}{2}$$

$$\text{Number of \{yellow, yellow\}} = NY \times \frac{NY - 1}{2}$$

$$\text{Number of \{red, blue\}} = NR \times NB$$

$$\text{Number of \{red, yellow\}} = NR \times NY$$

$$\text{Number of \{blue, yellow\}} = NB \times NY.$$

Note carefully the different "formula" for calculating the number of pairs of same colors and of different colors.

Note 5: Details of the counting of the different configurations may be found in Ben-Naim (2008). We present here only the final results (see Figs. 2.10, 2.11).

Note 6: Note again that we enclose "definition" in inverted commas. As we shall soon discuss, this is not a *bona-fide* definition. There are many stories about people who actually threw a die or a coin many times and counted the frequency of each outcome.

Note 7: It is almost impossible to decide between the two drugs. In both cases, the estimate of the probability of success is 60% (or 6/10). The second drug seems to be more "reliable" since it was tested on a much larger number of subjects. However, such an impression is unjustified. It is true (or rather, we believe that it is true) that the estimates of the *probabilities* are more reliable in case B than in case A. In other words, our extent of belief in the *probability* of success is larger in case B. This is not the same as the probability of success of the two drugs.

To put it differently, if the doctor is going to use the drugs on 1000 subjects, the doctor cannot tell which drug will be more successful than the other. All the doctor knows is that it is more *likely* that 60% of the patients will recover using drug B, and it is less certain that 60% will recover using drug A (it could be a higher or lower percentage). Thus, more experiments give you a better *estimate* of the *probabilities*. This does not mean *higher* probabilities or larger chances of success!

Note 8: Clearly, the overall statistics are in favor of drug C. This is consistent with the information provided by the drug company. Drug C has an 80% chance of success, while drug D only has a 70% chance of success, so without any further information, I would choose drug C.

However, when I got the additional information, I might or might not change my mind.

It seems that drug D works better for men than for women. Therefore, although the overall statistics points to a higher rate of success of drug C, I would choose drug D if I were a man.

If I were a woman, I would choose drug C. I would choose drug C as it seems to work better for women, as well as better in the total population.

Note also that no choice guarantees success. Furthermore, the result of using a drug is neither a black nor a white case. There is a huge spectrum of grey regions between success and failure. There are many degrees of recoveries. The drugs may be efficient for one person, but not for another, and might have different kinds of side effects. All of these factors should be taken into account when making a choice of using one drug over another.

Note 9: The probability is $\left(\frac{1}{10}\right)^{10}$, which is the same for each sequence. You will understand why after you read Session 4.

Session 3

Note 1: The sample space in case (i) is $\Omega = \{$red, blue, green, yellow, brown, black, white, purple$\}$.

The sample space in case (ii) can be chosen as either $\Omega = $ {red, blue, green}, or, if we label each of the faces by a number from 1 to 6, then $\Omega = \{1, 2, 3, 4, 5, 6\}$.

In case (iii), there are again several possibilities to define the sample space. One possibility is $\Omega = $ {marble, coin, dice}. Another possibility is to label each marble with a number (1 or 2), each coin with a number (1, 2 or 3) and each dice with a number (1, 2 or 3). The sample space in this case is $\Omega = $ {marble 1, marble 2, coin 1, coin 2, coin 3, dice 1, dice 2, dice 3}.

In case (iv), we can label the small squares with numbers from 1 to 100, and so the sample space is $\Omega = \{1, 2, 3, \ldots, 100\}$.

The same answer applies to case (v).

As you can see, for the same experiment, we can choose the sample space in different ways. There is no unique way of defining the sample space of any experiment.

Note 2: There is one impossible event, five events consisting of one elementary event, 10 events consisting of two elementary events, 10 events consisting of three elementary events, five events consisting of four elementary events and one certain event. The counts of each type of events are:

$$1, 5, 10, 10, 5, 1.$$

The total number of events in L is now 32.

Note 3: The result is $6 \times 5 = \frac{6!}{4!} = 30$. Figure 3.6 shows all of the possible arrangements.

Note 4: The binomial theorem is written as:

$$(X + Y)^n = \sum_{k=0}^{n} \binom{n}{k} X^k Y^{n-k}.$$

You can easily derive Newton's binomial theorem for a few small n, say:

For $n = 0$: $(X + Y)^0 = 1$.

For $n = 1$: $(X + Y)^1 = X + Y$.

For $n = 2$: $(X + Y)^2 = X^2 + 2XY + Y^2$.

For $n = 3$: $(X + Y)^3 = X^3 + 3X^2Y + 3XY^2 + Y^3$.

The particular case we discussed in this session is for the case $X = Y = 1$, i.e. $(1 + 1)^n$. The quantities $\binom{n}{k}$ are also called the binomial coefficients. These numbers appear in the general expansion of $(X + Y)^n$. They also appear in the Pascal triangle.

Note 5: Clearly, if the outcomes of throwing the dice are equally probable, event B has a larger probability than event A, and event D has a larger probability than event C.

Note 6: The results are: $Pr(X) = \frac{4}{6} = \frac{2}{3}$, $Pr(Y) = \frac{3}{6} = \frac{1}{2}$ and $Pr(X \cup Y) = Pr(1, 2, 3, 4, 6) = \frac{5}{6}$.

Note 7: Figure 3.10a shows three events A, B and C. Each of the pairs A and B, B and C, and A and C are disjoint as pairs. Also, the triplet A, B and C is disjoint as a triplet. Figure 3.10b shows an example in which the triplet A, B and C is disjoint as a triplet, i.e. the three areas A, B and C have no common overlapping area, $A \cap B \cap C = \emptyset$, but each pair is not disjoint. In general, if one pair is disjoint, then the triplet is also disjoint. On the other hand, if the triplet is disjoint, it does not follow that any of the pairs is disjoint. In Fig. 3.10c, we see three events that have a common area, which is $A \cap B \cap C \neq \emptyset$.

Note 8: See Fischbein and Schnarch (1997).

Session 4

Note 1: The answer to (a) is 1/2 for the first urn and 1/2 for the second urn. The probability of obtaining a red marble from the left urn *and* a red marble from the right urn is 1/4. The reason is that there are altogether eight marbles in the two urns. There are 16 possible pairs of one marble from the left and one from the right. Out of these 16 events, there are four events in which the marble drawn from the left urn is red and the marble drawn from the right urn is also red. Therefore, the probability of the required event is $4/16 = 1/4$.

The answer to (b) is also easy. There is a probability of 1/2 of picking a red marble on the first trial. There is also a probability of 1/2 of picking a red marble on the second trial (simply because the content of the urn does not change after the first draw).

The answer to (c) requires a little thought, but not much. Again, the probability of drawing a red marble in the first trial is 1/2. However, on the second trial, the "initial conditions" are different. You drew a red marble, but did not return it. So there are now one red marble and two blue marbles. Therefore, the probability of drawing a red marble on the second trial is 1/3. We say that the *conditional* probability of drawing a red marble, *given* that a red marble was already drawn (and not returned), is 1/3.

Note 2: Since the dice are stuck together, we expect the interaction between the two magnets to be strongest when the two magnets are in opposite directions as in Fig. 4.5. The event $(1, 4)$ will be very rare. In addition, the event $(6, 2)$ should be very rare. In this case, the correlation is expected to be very large and *negative*. On the other hand, the event $(1, 6)$ will occur very frequently. The event $(4, 4)$ will also occur with higher frequency than what we would expect from independent dice, hence the correlation is positive.

Note 3: From the first set of experiments, we can conclude that the two coins are fair or nearly fair. The results are nearly equal to each other and nearly equal to the product of the probabilities of each result on each coin. This means that the outcomes on the two coins are independent.

In the second run, it is clear that there is strong bias towards the events (H, H) or (T, T). I would suspect that there is a strong force acting between the two coins such that the H face of one is stuck to the T face of the other. Therefore, when they fall together, if one shows T upwards, the second will almost always show T upwards, and the same for H facing upward (Fig. 4.9). (Of course, you must separate the two coins to see the outcome of the one underneath the second.)

Note 4: If the marbles are completely independent, one would expect the probabilities of the four events to be:

$$Pr(a) = \frac{1}{4}, \quad Pr(b) = \frac{1}{4}, \quad Pr(c) = \frac{1}{4}, \quad Pr(d) = \frac{1}{4}.$$

The actual results of this experiment indicate that there is some attraction between the two marbles, e.g. one has a positive and one has a negative charge such that they attract each other. From the collected data, we can estimate the correlations between the two marbles as follows:

For event (a):

$$Pr(\text{red on left } and \text{ blue on left})$$
$$-Pr(\text{red on left}) \times Pr(\text{blue on left})$$
$$= \frac{40}{100} - \frac{50}{100} \times \frac{50}{100} = \frac{40}{100} - \frac{25}{100} > 0.$$

Note that we write 50/100 instead of 1/2 to simplify the calculations.

In this case, there is a *positive* correlation between the locations of the two marbles. The same is true for event (b). However, for cases (c) and (d) we find negative correlations.

For event (c):

Pr(red on the right *and* blue on the left)

$-Pr$(red on the right) $\times$ Pr(blue on the left)

$$= \frac{10}{100} - \frac{50}{100} \times \frac{50}{100} = \frac{10}{100} - \frac{25}{100} < 0.$$

We obtain a similar result for case (d).

In these cases, we have negative correlations between the locations of the two marbles.

Note 5: Clearly, as the marbles tend to be far away from each other, something must be repelling one from the other. Perhaps they carry the same electric charge. The estimated correlations are:

For event (a): the correlation is negative.

$$\frac{5}{100} - \frac{50}{100} \times \frac{50}{100} = \frac{5}{100} - \frac{25}{100} < 0.$$

A similar correlation is observed for event (b).

However, for case (c) we have a positive correlation between the locations of the two marbles.

$$\frac{45}{100} - \frac{50}{100} \times \frac{50}{100} = \frac{45}{100} - \frac{25}{100} > 0.$$

A similar positive correlation is observed for case (d).

Note 6: It seems that there is a strong attraction between the two marbles, as well as there being something about the left compartment that attracts the marbles. It could be a strong electrical or magnetic force, or perhaps centrifugal force giving preference to the marbles to be on the left side.

Note 7: From the definitions of COR and g, we have:

$$g(A, B) = \frac{Pr(A \cap B)}{Pr(A) \times Pr(B)} = \frac{COR(A, B) + Pr(A) \times Pr(B)}{Pr(A) \times Pr(B)}$$

or, equivalently:

$$g(A, B) = 1 + \frac{COR(A, B)}{Pr(A) \times Pr(B)}.$$

When $COR = 0$, $g = 1$;
When $COR > 0$, $g > 1$;
When $COR < 0$, $g < 1$.

Note 8: Your answer to question (1) was probably 1/6. You might have given the same answer to questions (2) and (3), i.e. $Pr(A) = 1/6$. However, if you did not blurt out your answers automatically, you may have wondered why I repeated the same, or almost the same trivial question three times? By the time you reached question (4), you must have been questioning the validity of the answers you gave to questions (1), (2) and (3).

Questions (1) to (4) were designed to prompt you to think about whether it is meaningful to assign a probability to any event without having *any additional information.*

You should realize that questions (1) to (3) are nearly the same, except that each time I added some more information. Usually, the additional information given in question (3) is taken for granted, and in giving your answer to questions (1) and (3), you implicitly assume that the same information given in (3) also applies to question (1).

However, strictly speaking, it is meaningless to assign a probability to an event without *any additional information*, much as it is meaningless to ask how long it takes to travel from A to B without providing some additional information on the distance between A and B, on the means by which you travel from A to B and perhaps on the time of the day you travel.

Usually, you assume that the information given *explicitly* in question (3) is also given *implicitly* in questions (1) and (2), and therefore you have correctly answered all three questions by saying $Pr(A) = 1/6$. However, when you read question (4), you may have realized that the information you had been assuming implicitly had changed. The answer to question (4) given the *new information* is $Pr(A) \approx 0$, because whenever you throw the dice, it will be pulled back by the spring to the same spot and to the same orientation as it was before it was thrown. The answer to question (5) is also $Pr(A) \approx 0$. If the metal side is very heavy, then most of the time the dice will fall in such a way as to show the face with one dot, as in Fig. 4.12.

Note 9: The probabilities given above can be determined by measuring the concentration of molecules having site a occupied, site b occupied and both sides occupied. We see that the probability of A is nearly equal to the probability of B. But the two events are not independent, i.e.:

$$Pr(A \cap B) \neq Pr(A) \times Pr(B) = \frac{1}{12} \times \frac{1}{10} = \frac{1}{120}.$$

We can calculate the correlation between the two events as follows:

$$g(A, B) = \frac{Pr(A \cap B)}{Pr(A) \times Pr(B)} = \frac{\frac{1}{200}}{\frac{1}{12} \times \frac{1}{10}} = \frac{12}{20} < 1$$

which means that the two events are negatively correlated.

Note 10: In the example of the two molecules in Fig. 4.19, it is easy to see that in the case of maleic acid (a), the two binding sites are closer than in the case of fumaric acid (b). As in the case of the two dice with magnets, we also expect there to be is a negative correlation between the two events: "binding of a proton (H^+) on the left side" and "binding of a proton on the right side" in both of the molecules g <1. The reason for this negative correlation is the repulsion between the positive charges on the two protons. In terms of conditional probabilities, we can say that the conditional probability of binding on the *second* site, given that the first site is already occupied, is *smaller* (because of the repulsion) than the probability of binding to a single site (the two sites are identical, so when we say the binding to the first site we mean binding to an unoccupied site when the other site is also unoccupied). Experiments confirm that, indeed, in the case of maleic acid, the correlation is more negative than in the case of fumaric acid. It is also known experimentally that if one measures the correlation of binding to two sites that are very far apart, the correlation becomes weaker and weaker as the distance between the two sites become larger.

Note 11: The interested reader is referred to Ben-Naim (2014) for details of this phenomenon.

Note 12: In all of these problems, my choice of a sequence is fixed: $\{1, 2, 3, 4, 5, 6\}$, and it has the probability of 1/2 of winning. If you choose the *disjoint* event $\{7, 8, 9, 10, 11, 12\}$, the conditional probability is (Fig. 4.21):

$$Pr(B|A) = Pr(7, 8, 9, 10, 11, 12|A) = 0$$

since knowing A in this case excludes the occurrence of B. In the next example, there is an overlap between A and B, hence:

$$Pr(B|A) = Pr(6, 7, 8, 9, 10, 11|A) = \frac{1}{6} < \frac{1}{2}.$$

Knowing that A occurred means that your winning is possible only if the ball landed on "6," hence the conditional probability is 1/6, which is smaller than $Pr(B) = 1/2$, i.e. there is a *negative correlation*. Note that in this game, both players can win. For instance, in the previous case, if the result is "6," then both players are winners.

Similarly, for the next example:

$$Pr(B|A) = Pr(5, 6, 7, 8, 9, 10|A) = \frac{2}{6} < \frac{1}{2}.$$

Here, "given A," you win only if the ball lands on either "5" or "6," hence the conditional probability is 2/6, which is still smaller than $Pr(B) = 1/2$.

In the next case:

$$Pr(B|A) = Pr(4, 5, 6, 7, 8, 9|A) = \frac{3}{6} = \frac{1}{2}.$$

Here, the conditional probability is 1/2, which is exactly the same as the "unconditional" probability $Pr(B) = 1/2$, meaning that the two events are *independent*.

Next we have:

$$Pr(B|A) = Pr(3, 4, 5, 6, 7, 8|A) = \frac{4}{6} > \frac{1}{2}$$

$$Pr(B|A) = Pr(2, 3, 4, 5, 6, 7|A) = \frac{5}{6} > \frac{1}{2}$$

$$Pr(B|A) = Pr(1, 2, 3, 4, 5, 6|A) = \frac{6}{6} = 1 > \frac{1}{2}.$$

In the last example, knowing that A occurs makes the occurrence of B *certain*. In these examples, we have seen that overlapping events can either be positively correlated, negatively correlated or non-correlated.

Note how the correlation changes from extremely negative ("given A" *certainly* excludes your winning in the first example) to extremely positive ("given A" assures you of winning in the last example). At some intermediate stage, there is a choice of a sequence that is *indifferent* to the information "given A." What is this choice?

Note 13: Assume for simplicity that the total area of the board is *unity*. Then the probability of hitting A is simply the area of A. The probability of hitting B is the area of B. Let $p = Pr(A) = Pr(B)$. The conditional probability is calculated by the rule $Pr(A|B) = \frac{Pr(A \cap B)}{Pr(B)}$, where $Pr(A \cap B)$ is the probability of hitting both A *and* B. Denote this probability by x (this is equal to the overlapping areas of A and B). The correlation between the two events "hitting A" and "hitting B" is defined as $g = \frac{Pr(A \cap B)}{Pr(A)Pr(B)} = \frac{x}{p^2}$. When the overlapping area is $x = 0$, then $g = 0$ (i.e. extreme negative correlation). When the overlapping area is $x = p$, we have $g = p/p^2 = 1/p$ or $Pr(A|B) = \frac{p}{p} = 1$ (i.e. extreme positive correlation). When $x = p^2$, we have $g = p^2/p^2 = 1$ or $Pr(A|B) = Pr(A)$; in this case, there is no correlation between the two events, i.e. the two events are *independent*. For the particular example in Fig. 4.22, $p = 1/10$, therefore at $x = 1/100$, we have independence of the two events.

Note 14: Sometimes the results found by Falk (1979) are referred to as the "Falk phenomenon."

Note 15: Note that A and B have a large intersection area, hence A supports B (as well as B supports A). Similarly, B and C have a large intersection area, hence B supports C (as well as C supports B). See Fig. 4.22 in order to refresh your memory about the extent of overlapping areas and the sign of the correlation between two events. Note also that neither A, B nor C are certain events. See also Note 13.

It is clear that A and C do not have an intersection area; therefore, A does not support C.

Note 16: Initially, the judge could have reached some conclusion regarding the probability that the suspect was guilty. Let us say that she assumed that the probability of the suspect's guilt *given* the suspect's past criminal record was about 50%. We write this in shorthand notation as:

$$Pr(\text{guilt}|\text{past record}) \approx 50\%.$$

After reading the evidence presented by Chief Greenhead, the judge could modify her estimate upwards, i.e. the new evidence supports the defendant's guilt:

$$Pr(\text{guilt}|\text{past record } and \text{ Greenhead evidence}) \approx 80\%$$
$$> Pr(\text{guilt}|\text{past record}).$$

The same conclusion could be reached from Officer Redhead's evidence:

$$Pr(\text{guilt}|\text{past record } and \text{ Redhead evidence}) \approx 80\%$$
$$> Pr(\text{guilt}|\text{past record}).$$

We can say that each of the officer's evidence reckoned *by itself* supports the contention of the defendant's guilt. However, *taken together*, the two pieces of evidence exonerate the suspect:

$$Pr(\text{guilt}|\text{past record } and \text{ Greenhead } and \text{ Redhead evidence})$$
$$\approx 0.$$

Clearly, if it is known that the suspect spent about 1 h 25 min in the restaurant, and left the restaurant at 12.25 a.m., and if it also known that he entered the bar at 12.35 a.m. and stayed there for 1 h 25 min, he could not have possibly spent the full 30 min in the Victimberg's residence (see diagram in Fig. 4.25). Thus, although *each* of the pieces of evidence is supportive, the combined evidence is unsupportive.

Figure 4.26 shows an example where B supports A and C supports A, but $B \cap C$ (B *and* C) does not support A. Refresh your memory about the relationship between the extent of overlapping areas and the extent of dependence (see Fig. 4.22).

Figure 4.27 shows an example where B does not support A. Furthermore, C does not support A, but the intersection B and C supports A. In fact, $B \cap C$ actually makes A certain.

The story of the two types of incriminating evidence is an example of the case that is shown schematically in Fig. 4.26. Can you invent a story that is relevant to Fig. 4.27?[17]

Note 17: Suppose a crime was committed in region A, as in Fig. 4.27. A suspect claimed that on the day when the crime was committed, he can prove that he was in region B. This means that the evidence "being in B" does not support that he was in A. The next day, a policeman reported that he saw the suspect in region C on the day of the crime. This evidence also does not support that he was in A. However, both of the evidence B and C support the claim that he *was* in A on the day of the crime.

Note 18: We know that:

$$Pr(\mathrm{B}) = Pr(\mathrm{R}) = Pr(\mathrm{G}) = \frac{1}{2}.$$

We also know that:

$$Pr(\mathrm{B} \cap \mathrm{R}) = \frac{1}{4} = Pr(\mathrm{B}) \times Pr(\mathrm{R}) = \frac{1}{4}$$

$$Pr(\mathrm{B} \cap \mathrm{G}) = \frac{1}{4} = Pr(\mathrm{B}) \times Pr(\mathrm{G}) = \frac{1}{4}$$

$$Pr(\mathrm{R} \cap \mathrm{G}) = \frac{1}{4} = Pr(\mathrm{R}) \times Pr(\mathrm{G}) = \frac{1}{4}.$$

However:

$$Pr(\mathrm{B} \cap \mathrm{R} \cap \mathrm{G}) = \frac{1}{4} \neq Pr(\mathrm{B}) \times Pr(\mathrm{R}) \times Pr(\mathrm{G}) = \frac{1}{8}.$$

Therefore, in this example, there is independence in pairs, but not in triplets.

Note 19: This problem was actually given to students. Here are two possible answers, as well as their reasons.

The first answer: I expect that the result will be H. The reason is that if the coin has shown the same outcome H 1000 times in a row, the coin must be biased towards falling with its T face downwards. Therefore, I expect that the 1001th trial will result in an H too!

The second answer: I expect that the next result will be T. I know that the coin is fair, so the probability of each result is 1/2. Since there were so many outcomes of H, it is time that a few or perhaps more than a few Ts should appear to change the balance of events. In other words, I expect that from now on, more Ts will appear, such that the total number of Ts and Hs will be balanced. Therefore, I expect that the next 1001th trial will be T!

Both answers are very logical. There is nothing wrong with the reasoning of the two answers. However, those who gave the first answer ignored the *given* (and emphasized) *fairness* of the coin. Given that the coin is fair, you cannot assume that it might be biased. Those who gave the second answer ignored the *given* (and emphasized) *independence* of the events. Thus, the logic of the reasoning is correct, but in both cases the persons who gave the answers ignored the *given* conditions. If you read the given conditions again, then you must conclude that:

$$Pr(H|given\ one\ thousand\ Hs\ in\ a\ row) = \frac{1}{2}.$$

This example is important. Sometimes people confuse the probability of the joint events (A *and* B) with the conditional probability (A|B). The 1000 Hs in a row is indeed a very rare event. It has a very small probability of occurring. However, the conditional probability that the next result will be H (or T) *given* that H occurred 1000 times is still 1/2 (provided that the coin is fair and the throws are independent).

Note 20: We shall learn the theorem of total probability in the next session. Here, we use just common sense to solve the problem. The solution to the problem is this. Denote by X the probability of Linda winning. Clearly, in the next throw after the game is halted, there are three mutually exclusive cases with probabilities:

(I)   Outcome {6} with probability 1/6;
(II) Outcome {4} with probability 1/6;
(III) Outcome {1, 2, 3, 5} with probability 4/6.

Let us denote the event "Linda wins" by *LW*. Using the theorem of total probability, the following equation holds:

(IV) $X = Pr(LW) = Pr(I)Pr(LW|I) + Pr(II)Pr(LW|II)$

$$+ Pr(LW)Pr(LW|III)$$

$$= \frac{1}{6} \times 1 + \frac{1}{6} \times \frac{1}{2} + \frac{4}{6} \times X.$$

This is an equation with one unknown: $6X = \frac{3}{2} + 4X$. The solution is $X = 3/4$. Note that the events (I), (II) and (III) refer to the possible outcomes on the *next* throw. However, the event *LW* refers to *Linda wins*, regardless of the number of subsequent throws. Equation (IV) means that the probability of Linda winning is the sum of the three probabilities of the three mutually exclusive events. If (I) occurs, then she wins with probability one. If (II) occurs, then she has a probability 1/2 of winning (since in this case both will have two points). If (III) occurs, then the probability of her winning is X, i.e. it is the same as at the moment of halting the game. Therefore, the two players should divide the total sum in such a way that Linda gets 3/4 and Dan gets 1/4.

Note 21: Sorry, I cannot give you the probabilistic answer to this question. If the area of the square is about 100 m^2, the length of the diagonal is about $10\sqrt{2}$ m and the width of the "safe strip" is 10 cm, then the area of the safe strip is about $\sqrt{2} \approx 1.4$ m^2.

Therefore, if I knew where the square was, and I jumped into said square, the probability of landing on the safe strip is about 1.4/100. But since I do not know where the square is, the probability of landing on that strip is much smaller. Even if I landed on a safe spot (which could be either on the safe strip or outside the square), I still would not know in which direction to walk in order to get to the center of the square.

Since this book is about probability, and since there exists no probabilistic answer to this question, I could end this note at this point. However, if you are curious as to why I bothered you with this unsolvable problem, let me tell you that while I admit that there is no probabilistic solution to the problem, perhaps you would be happy to learn that there is an *exact* geometric solution to this problem. Figure 4.32 shows the original square on the left-hand side, and what remains; the four stakes, A, B, C and D, on the right-hand side. The geometrical problem is the following: Given the four points A, B, C and D, how do we reconstruct the original square from these four points? The procedure of reconstructing the original edges of the square is described in Fig. 4.32 to Fig. 4.35. If you are keen to get a probabilistic answer, I can reformulate my answer as follows:

The probability of finding the treasure by searching at random is very small. However, the *conditional probability* of finding the safe strip, and hence the treasure, given that you know geometry, turns into one!

The moral: It is worth investing in studying geometry before walking into a field full of landmines.

Session 5

Note 1: The solution to this problem is exactly the same as we have found for the dart hitting the board. Mathematicians would say that the two problems are isomorphic except for the notation in this example, in which D replaces B from the previous example.

You have all the *a priori* probabilities:

$$Pr(A_1) = \frac{1}{4}, \quad Pr(A_2) = \frac{3}{4},$$

$$Pr(D|A_1) = \frac{1}{10}, \quad Pr(D|A_2) = \frac{2}{10},$$

$$Pr(D) = \frac{1}{10} \times \frac{1}{4} + \frac{2}{10} \times \frac{3}{10} = \frac{7}{40}.$$

Hence the required probabilities are:

$$Pr(A_1|D) = \frac{Pr(A_1 \cap D)}{Pr(D)} = \frac{Pr(D|A_1)Pr(A_1)}{Pr(D)}$$

$$= \frac{\frac{1}{10} \times \frac{1}{4}}{\frac{7}{40}} = \frac{1}{7}$$

$$Pr(A_2|D) = \frac{Pr(A_2 \cap D)}{Pr(D)} = \frac{Pr(D|A_2)Pr(A_2)}{Pr(D)}$$

$$= \frac{\frac{2}{10} \times \frac{3}{4}}{\frac{7}{40}} = \frac{6}{7}.$$

Note 2: By now you should know how to dissuade the poor guy from the story at the beginning of this session from committing suicide.

Note 3: Here, we saw how the *same answer* given to two different persons carries two different messages. The story below shows how the *same question*, when addressed to two different persons, has different meanings.

How to Tell the Difference Between an Honest and a Pseudo-Honest Person

This puzzle is adapted from Mero (1990).

Not far from the Amazon region lie lush and verdant hills and valleys. At the foot of one of these valleys lies a village inhabited by only two types of people: The honest and the pseudo-honest. The honest people always see the world as it actually is, and they never lie. The pseudo-honest people, on the other hand, see the exact opposite of what reality is, and they never tell the truth. So they will give you an answer that is the opposite of what they see. When asking an honest person if the grass is green, the answer will be "Yes'. If you ask a pseudo-honest person the *same* question, he or she will see that the grass is not green, but will, however, lie and say "Yes." Both the honest and the pseudo-honest persons will always give the *same* answer to the *same* questions which require a Yes or a No answer. Therefore, given this scenario, differentiating the honest from the pseudo-honest persons is *impossible* in this village.

You go to the village for some business prospects and meet two of its inhabitants, John and Jack. You know that they come from the same village, but you do not know whether John or Jack are honest or pseudo-honest. In the pretext of making friends, you talk to them and probe deeper, but much to your chagrin, you realize that nothing in their behavior gives away their type of personality.

Is there any way to find out who is who in spite of the fact that they will always give the *same* answer to the *same* question?

Yes, you can easily check that by asking John, "Are you honest?" If he is honest, he will say "Yes"; if he is pseudo-honest, he perceives himself as honest, but he will lie and answer with a "No." The same applies to Jack, who will answer "Yes" if he is honest and "No" if he is pseudo-honest.

Therefore, by asking anyone of them "are you honest?" you will know who is who.

This, however, seems to contradict the conclusion reached before that the two persons of different types will always give the *same* answer to the *same* question. So, how come they gave different answers to the *same* question?

The answer is simple. The two questions are not the same. The *words* are the same — "Are you honest?" — but the "you" in the two questions are different. In other words, your question to John is, "Are you, John, honest?" This is different from the question addressed to Jack: "Are you, Jack, honest?"

Another solution to the apparent paradox is to change the conclusion that each person of the village gives the *same* answer to the *same* question. Instead, we know from the description of the behavior of the persons of the village that each person says the *truth* to each question.

For instance, if you ask, "Is 5 equal to 5?" the honest person's answer will be "Yes." The pseudo-honest person will also answer "Yes." If you ask, "Is 5 smaller than 2?" clearly the honest person will say "No," and the pseudo-honest person will say "No" as well.

From this conclusion, it follows that if you ask the honest person, "Are you honest?" they will say the truth — "Yes." If you ask the pseudo-honest person, "Are you honest?", obviously, they will also say the truth — "No."

Note 4: For details, see Ben-Naim (2008).

Note 5: The answer is:

$$Pr(W_2 | W_1) = \frac{N - 1}{N + N - 1}.$$

The reasoning is simple. After we drew a white (W_1), there are $N - 1$ white balls, and a total of $N + (N - 1)$ balls in the urn.

The inverse question can be derived from the identity:

$$\frac{Pr(W_2 | W_1)}{Pr(W_1 | W_2)} = \frac{Pr(W_2 \cap W_1)}{Pr(W_1)} \times \frac{Pr(W_2)}{Pr(W_2 \cap W_1)} = \frac{Pr(W_2)}{Pr(W_1)}.$$

Since $Pr(W_1) = Pr(W_2)$, the two conditional probabilities on the left-hand side must also be equal. Remember that $Pr(W_1)$ is the probability that a white ball is drawn on the first draw.

Likewise, $Pr(W_2)$ is the probability that a white ball is drawn on the second draw. Each of these probabilities is equal to 1/2. I hope that you can appreciate the simplicity, elegance and beauty of this solution.

Session 6

Note 1: Gary's argument is indeed very powerful. Seatbelts can be dangerous in some accidents. However, those accidents are extremely rare. Therefore, on *average*, it is safer to use seatbelts instead of not using them.

Note 2: Even without having a definition of an *average* quantity, it is clear that the statement made by ASU in 2000 is meaningless. The grades of *all* of the students cannot be *above* the average. What the writer of this statement probably wanted to say is that the average grade in 2000 was above the averages of the preceding decades, or perhaps above the average of all students in the country. The *average* grade of the students in a given university is a number between the lowest and the highest grade.

Note 3: Yes, it is possible. It is easy to construct an example. Suppose HSU has seven professors with the following IQs: 100, 110, 120, 130, 140, 150 and 160; the average is 130. LSU has also seven professors with IQs: 50, 60, 70, 80, 90, 100 and 110; the average is 80.

Now, if the professor of the lowest IQ from HSU moves to LSU, the new situation is:

In HSU, there are six professors with IQs 110, 120, 130, 140, 150 and 160.

In LSU, there are eight professors with IQs 50, 60, 70, 80, 90, 100, 100 and 110.

Check that the average IQ in HSU *increased* from 130 to 135, and also the average IQ in LSU *increased* from 80 to 82.5.

Of course you cannot increase the average IQ of *all* of the professors in the two universities. It was 105 before the move and 105 after the move. Explain why.

Note 4: Note that the average of the *two speeds* is $\frac{140}{2} = 70$ km/h. However, in order to calculate the average speed on the round trip, you take the total length and divide it by the total time of travel. In our examples, let the distance between Jerusalem to Tel Aviv be a. The travel time to Tel Aviv is $t_1 = \frac{a}{v_1} = \frac{a}{40}$, and the travel time from Tel Aviv to Jerusalem is $t_2 = \frac{a}{v_2} = \frac{a}{100}$. Therefore, the *average speed* in the round trip is $\frac{2a}{t_1+t_2} = \frac{2a}{\frac{a}{40}+\frac{a}{100}} = \frac{8000}{140} \approx 57$ km/h.

In general, if the first speed is v_1 and the second is v_2, then the average of the two speeds is always *larger* than the average speed on the round trip, i.e.:

$$\frac{v_1 + v_2}{2} - \frac{2v_1 v_2}{v_1 + v_2} = \frac{(v_1 - v_2)^2}{2(v_1 + v_2)} \geq 0.$$

Note 5: The average of the two speeds is approximately *half* the *speed of light*. The average speed of the entire round trip is *half* the *donkey's speed*.

Note 6: The average of the two speeds is $\frac{(v+w)+(v-w)}{2} = v$. This is independent of w. However, the average speed in the round trip journey is calculated as follows:

Suppose that the distance from Tel Aviv to New York is x. The total flight time is:

$$t_1 + t_2 = \frac{x}{v - w} + \frac{x}{v + w} = \frac{2vx}{v^2 - w^2}$$

and the average speed in the round trip is:

$$\frac{2x}{t_1 + t_2} = \frac{v^2 - w^2}{v}.$$

When $w = 0$ (today), the average speed is the same as the average of the two speeds, i.e. v. For any other w, which is smaller than v ($w < v$), the average speed is smaller than v. Hence, the total time of the round trip flight is *longer* (tomorrow) than when there is no wind (today). So you had better fly today!

If the wind is too strong, say $w = v$, then the average speed will be zero and the flight time will be *infinity*. If the wind speed is even larger than v, then you will never get to New York!

Note 7: Note that in Fig. 6.5 we plot the *distribution density*. $f(v)dv$ is the probability that a molecule will have a speed between v and $v + dv$.

Note 8: This result follows from the very definition of the average quantity. For any average of the variable X, we have:

$$\sum_T Pr(X)(X - \bar{X}) = \sum Pr(X)X - \bar{X} \sum Pr(X) = \bar{X} - \bar{X} = 0$$

Note that we used the "normalization" condition $\sum Pr(x) = 1$.

This result might be surprising at first glance, and it might mislead you into concluding that the width of the curve is zero. However, a little thought will clarify this apparently surprising result. We took the average value of the *distance* from $\bar{T}$. Clearly, some distances are *positive* (for those temperatures higher than $\bar{T}$), and some distances are *negative* (for those temperatures lower than $\bar{T}$), and therefore, *on average*, the sum over the positive values will cancel the sum over the negative values, and we get zero. Note that this result is true for any distribution. In some textbooks, you might find that this statement is valid only for symmetric distributions.

Session 7

Note 1: The probability of finding a specific marble (any *specific* marble) in a specific cell is 1/10. The reason is that we assumed that all of the cells are equivalent and the probability of a specific marble being in a specific cell is independent of the number or the color of the marble.

The most probable distribution of marbles in the cells is the uniform distribution. Note again the term "uniform distribution" here means an equal probability of finding a marble in any one of the cells. The most probable distribution refers to the probability of the *entire* distribution.

You can also try to do the experiment on a computer. Instead of shaking the box, you can generate random numbers. You will get the same eventual distribution. This is quite a remarkable result. From a totally random process in which you cannot predict the location of any marble, you get a precise and predictable *final distribution*. In this particular experiment, the most probable distribution is the uniform distribution of the marbles in the cells. The same principle holds for the distribution of molecules in a room. For an actual simulation, see Ben-Naim (2008).

Note 2: For a *specific* sequence, the probability is $p^3 q^3$. We count 20 such specific sequences. Therefore, the probability of *any* sequence with three Hs and three Ts is $20p^3 q^3$.

Note 3: The numerators in these ratios are given in the Pascal triangle.

Note 4: The probabilities are (n is the number of particles in the left-hand compartment):

$$Pr(0,6) = \left(\frac{2}{3}\right)^6$$

$$Pr(1,5) = 6\left(\frac{1}{3}\right)\left(\frac{2}{3}\right)^5$$

$$Pr(2,4) = 15\left(\frac{1}{3}\right)^2\left(\frac{2}{3}\right)^4$$

$$Pr(3,3) = 20\left(\frac{1}{3}\right)^3\left(\frac{2}{3}\right)^3$$

$$Pr(4,2) = 15\left(\frac{1}{3}\right)^4\left(\frac{2}{3}\right)^2$$

$$Pr(5,1) = 6\left(\frac{1}{3}\right)^5\left(\frac{2}{3}\right)$$

$$Pr(6,1) = \left(\frac{1}{3}\right)^6.$$

Note 5: For simulated experiments of this kind, see Ben-Naim (2007, 2010).

Note 6: For details, see Ben-Naim (2010, 2012)

Note 7: The normal function was probably first used by the mathematician Abraham de Moivre (1667–1754). The mathematical details of this function were worked out by Gauss.

Note 8: For a derivation, see Ben-Naim (2008).

Session 8

Note 1: The interested reader is advised to consult Ben-Naim (2008, 2012).

Note 2: If you have any difficulty with logarithms, remember that the logarithm of any number (x) to the base 2 is the number that you have to put at the power of 2, i.e. $2^{(-)}$, so that the result is x. For example, $\log_2 2 = 1$ since $2^1 = 2$. Similarly, $\log_2 8 = 3$ since $2^3 = 8$. Also remember that $\log_2 xy = \log_2 x + \log_2 y$ and $\log_2 (1/2) = -1$. This is all that you need to know in mathematics regarding these matters.

These rules follow from the properties of the exponential. For any integers n and m:

$$2^n = 2 \times 2 \times 2 \times \cdots \times 2 \text{ (multiply 2 by itself } n \text{ times).}$$

Also:

$$2^{n+m} = (2 \times 2 \times 2 \times \cdots \times 2) \times (2 \times 2 \times 2 \times \cdots \times 2)$$
$$= 2^n \times 2^m$$

Denote $x = 2^n$ and $y = 2^m$; then $\log_2 xy = n + m = \log_2 x + \log_2 y$.

Similarly, since

$$2^{-n} = \frac{1}{2^n},$$

therefore:

$$\log_2 \left(\frac{1}{x}\right) = \log_2 \left(\frac{1}{2^n}\right) = -n = -\log_2 x.$$

We show here the rules for integer numbers n and m. However, the logarithm function can be used for any positive number.

Note 3: If you know differential calculus, you can easily prove that this function has a single maximum at $p = 1/2$.

Note 4: To prove that the SMI has a maximum for a specific distribution, we first define the auxiliary function:

$$F(p_1, p_2, \ldots, p_n) = -\sum_{i=1}^{n} p_i \log p_i + \lambda \left(\sum_{i=1}^{n} p_i - 1 \right).$$

The requirement for an extremum is that the derivative of F with respect to each p_i is zero, i.e.:

$$\left(\frac{\partial F}{\partial p_i} \right) = -\log p_i - 1 + \lambda = 0.$$

Hence:

$$p_i = 2^{(\lambda - 1)}.$$

This means that p_i is a constant, independent of the index i. To determine the value of p_i, we use the normalization condition:

$$1 = \sum_{i=1}^{n} p_i = \sum_{i=1}^{n} 2^{(\lambda - 1)} = n2^{(\lambda - 1)}.$$

Hence:

$$p_i = 2^{(\lambda - 1)} = \frac{1}{n} \quad \text{for } i = 1, 2, \ldots, n.$$

Thus, the SMI has a maximum value for the specific *uniform* distribution $p_i = 1/n$. For more details, see Ben-Naim (2008).

Note 5: You can also play this game on the computer, choosing any N and any strategy you wish. See ariehbennaim.com→ books→ *Entropy Demystified*→ simulated games.

Note 6: For details, see Ben-Naim (2010, 2012).

Note 7: For details, see Ben-Naim (2008).

Note 8: Actually, the number of configurations is further reduced because of the indistinguishability of the atoms and molecules. For more details, see Ben-Naim (2008).

References and Suggested Reading

Alvarez, L. W. (1965). A pseudo experience in parapsychology. *Science* **148**: 1541.

Ben-Naim, A. (2001). *Cooperativity and regulation in biochemical systems.* New York, NY: Plenum Press.

Ben-Naim, A. (2007). *Entropy Demystified.* Singapore: World Scientific.

Ben-Naim, A. (2008). *A farewell to entropy, statistical thermodynamics based on information.* Singapore: World Scientific.

Ben-Naim, A. (2009). *Molecular theory of water and aqueous solutions, part I: Understanding water.* Singapore: World Scientific.

Ben-Naim, A. (2010). *Discover entropy and the second law of thermodynamics: A playful way of discovering a law of nature.* Singapore: World Scientific.

Ben-Naim, A. (2012). *Entropy and the second law interpretation and miss-interpretations.* Singapore: World Scientific.

Ben-Naim, A. (2014). *Statistical Thermodynamics with Applications to the Life Sciences.* World Scientific, Singapore.

Bennett, D. J. (1998). *Randomness.* Cambridge, MA: Harvard University Press.

Bent, H. A. (1965). *The second law.* New York, NY: Oxford University Press.

David, F. N. (1962). *Games, gods and gambling, a history of probability and statistical ideas.* New York, NY: Dover Publication.

Falk, R. (1975). *Perception of randomness.* Unpublished doctoral dissertation, The Hebrew University of Jerusalem, Israel

Falk, R. (1979). *Understanding probability and statistics. A book of problems.* Wellesley, MA: A K Peters.

Falk, R., Falk, R. and Levin, L. (1980). *A potential for learning probability in young children. Edu Studies Math* **11**: 181.

Falk, R. (1981–1982). On coincidence. *Skeptical Inquirer* **6**: 18–31.

Falk, R., Yudilevich-Assouline, P., & Elstein, A. (2012). Children's concept of probability as inferred from their binary choices — revisited. *Edu Studies Math* **81**: 207–233.

Fischbein, E., & Schnarch, D. (1997). The evolution with age of probabilistic, intuitively based misconceptions. *J Res Math Edu* **28**: 96–105.

Huff, D. (1952). *How to Lie with Statistics*. New York, NY: W. W. Norton & Company.

Mérö, L. (1990). *Ways of Thinking*. World Scientific, Singapore.

Moran, P. A. (1984). *An Introduction to Probability Theory*. Oxford, UK: Oxford University Press.

Unwin, S. (2004). *The Probability of God: A Simple Calculation that proves the Ultimate Truth*. Three Rivers Press.

vos Savant, M. (1972) *Brain Building. Exercising Yourself Smarter*. Bantam, New York.

Index